# Recommendations for the inspection, maintenance and management of car park structures

### National Steering Committee for the Inspection of Multi-storey Car Parks

Institution of Civil Engineers

Published by Thomas Telford Publishing, Thomas Telford Ltd, 1 Heron Quay, London E14 4JD.
www.thomastelford.com

First published 2002

Distributors for Thomas Telford books are
*USA:* ASCE Press, 1801 Alexander Bell Drive, Reston, VA 20191-4400, USA
*Japan:* Maruzen Co. Ltd, Book Department, 3–10 Nihonbashi 2-chome, Chuo-ku, Tokyo 103
*Australia:* DA Books and Journals, 648 Whitehorse Road, Mitcham 3132, Victoria

A catalogue record for this book is available from the British Library

ISBN 978-0-7277-3183-8

Typeset by Gray Publishing, Tunbridge Wells, Kent

# Foreword

There are well over 4000 multi-storey car parks in the UK. Almost all have been built since 1940 with a boom in their construction in the 1960s.

Most car park structures are of reinforced concrete construction. Many have a history of early deterioration, structural defects and safety shortcomings due to poor design and construction and low standards of maintenance and repair. They were designed to normal building standards. Experience has shown that they are subject to a much more severe environment, more akin to the exposure of bridges. As a result, deterioration, particularly reinforcement corrosion due to the effects of de-icing salt, has had a major impact on their durability. There have been a few partial collapses, and accidents involving vehicles breaking through edge protection barriers have also occurred. As a result closures of car parks for costly repairs and/or rehabilitation have been required. These experiences have emphasized the need for improved performance and safety of existing car park structures.

These Recommendations address important implications of the past performance of existing car park structures for their safety and use in the future. They provide guidance on standards of care that are considered to constitute good practice. The Recommendations have been prepared by the Institution's National Steering Committee for the Inspection of Multi-storey Car Parks.

Part 1 of the Recommendations is intended for use by Owners and Operators. Part 2 is intended for their professional engineering advisers. The Recommendations, as a whole, aim to assist both Owners and Operators and their professional advisers to provide safe car park structures and cost-effective inspection and maintenance with minimum disruption.

It is my pleasure to acknowledge the contributions made by members of the National Steering Committee, their advisers, the organizations that sponsored the work, the consultant to the Committee and the many members of the profession and industry who provided information and comment during the preparation of these Recommendations.

Adrian Long
President
Institution of Civil Engineers
December 2002

# Acknowledgements

These Recommendations have been prepared by the National Steering Committee for the Inspection of Multi-storey Car Parks of the Institution of Civil Engineers.

*Members*

| | |
|---|---|
| Mr David Yeoell (Chairman) | London Boroughs Technical Advisers Group |
| Mr Douglas Pittam (Secretary/Treasurer) | Consultant |
| Mr Barry Ayres (Deputy Chairman) | Association for Structural Engineers of London Boroughs |
| Mr Amar Bhogal | Institution of Civil Engineers |
| Ms Berenice Chan | Institution of Structural Engineers |
| Mr John Collins | Welsh Assembly Government |
| Dr Colin Jolly | Cranfield University – RMCS |
| Mr Jolyon Kenward | Hyder Consulting Limited |
| Mr William Lillis | National Car Parks Limited |
| Mr Steven Luke | Ove Arup and Partners |
| Mr Rick Mallett | RJ Witt Associates (Babtie Group Limited until August 2001) |
| Mr David Merriman | CSS |
| Mr Brian Neale | Health and Safety Executive |
| Professor Peter Robery | FaberMaunsell |
| Mr Tony Trezies | Jenkins and Potter |
| Mr Paul Watters | Automobile Association |

*Consultant to the Committee*

| | |
|---|---|
| Dr John Menzies | Consultant |

*The Committee would like to thank the following for their contributions:*

| | |
|---|---|
| Mr John Barber | |
| Mr Peter Beveridge | Stirling Maynard and Partners |
| Mr David Bigland | Building Research Establishment Ltd |
| Mr David Bravery | Mitchell McFarlane and Partners Ltd |
| Mr F Domaingue | London Borough of Richmond |
| Mr John Drewett | Concrete Repairs Ltd |

| Dr Mike Grantham | MG Associates |
| Mr Mike Haynes | FaberMaunsell |
| Dr Neil Henderson | Mott MacDonald |
| Mr Gordon Masterton | Babtie Group Limited |
| Dr Stuart Matthews | Building Research Establishment Ltd |
| Professor George Somerville | UK Cares |
| Mr Neil Telfer | Bullen Consultants Ltd |
| Mr RJ Walling | Mason Revis Partnership |
| Mr Chris Whapples | Hill Cannon Partnership |
| Dr Jonathan Wood | Structural Studies and Design Limited |

*Financial contributions by the following organizations to the preparation of these Recommendations are gratefully acknowledged:*

British Parking Association
National Car Parks Limited
London Boroughs Technical Advisory Group
Hill Cannon Partnership
RJ Witt Associates
Babtie Group Limited
Dundee City Council

*Photograph Acknowledgements*

Photographs have been supplied by courtesy of, and are published with the permission of the following organizations or individuals:

| Jenkins & Potter | Figure 3.1 |
| FaberMaunsell | Figure 13.2 |
| Glasgow City Council | Figures 2.2, 7.1, 15.1, 15.3 and 15.5 |
| Hill Cannon Partnership | Figures 12.3, 13.3 and 15.4 |
| Dr Colin Jolly | Figure 15.7 |
| National Car Parks Limited | cover, Figures 3.2 and 15.6 |
| Mott MacDonald | Figures 2.1, 6.1, 12.2 and 13.1 |

# Contents

# Glossary

**Assistant**  A person with knowledge of car park structures and the deterioration processes likely to be present who, through training, is competent to carry out Routine Inspections using a checklist under the supervision of the Engineer or the Inspector.

**Car park structure**  Primary framework essential to the support and stability of a multi-storey car park, including key elements, secondary elements, access ramps and bridges and retaining walls.

**Cladding**  External components fixed to the car park structure.

**Condition Survey**  The visual examination of all accessible areas of a car park structure and associated fabric/elements, and measurement of evidence of deterioration, including any Structural Investigation.

**Daily surveillance**  Visual observation and reporting of equipment breakdown, obvious deterioration and damage to the car park structure, cladding and edge protection, and of untoward incidents in the use of the car park.

**Edge protection**  A barrier or restraint around the edge of a car park deck to restrain errant vehicles (vehicle edge restraint barrier) and/or to protect pedestrians (pedestrian guarding).

**Engineer**  A chartered civil or structural engineer with appropriate experience to take the lead in advising an Owner/Operator on structural safety and on the management, inspection, maintenance and repair of a car park structure, cladding and edge protection.

**Initial Appraisal**  An overall structural and materials appraisal of a car park structure, cladding and edge protection based on a desk study of available records and factual information derived from Condition Survey.

**Inspection and maintenance strategy**  A strategy involving inspections and maintenance of a car park structure meeting the requirements of the Owner/Operator.

**Inspector**  A person competent by virtue of a combination of qualification, experience and training to undertake Condition Surveys of car park structures. Depending on organizational arrangements, the Inspector may be employed by the same or a different organization from the Engineer. The Inspector will usually be an experienced chartered civil or structural engineer.

**Key element**   A component of a structure whose failure would cause the collapse of more than a limited portion close to the component in question.

**Life-care Plan**   A long-term plan for the implementation of an inspection and maintenance strategy for a car park structure.

**Multi-storey car park**   An elevated, multi-storey car park, whether self-contained or forming part of a multi-use structure, including any storeys below ground and any bridging or elevated elements essential to the car park.

**Owner and Operator**   The Owner is the owner of the multi-storey car park. The Operator is the organization responsible for the operation and use of the car park. The Operator may be the Owner or a parking operations contractor.

**Rehabilitation**   The repair and refurbishment of a part or the whole of a car park structure.

**Repair**   Action taken to reinstate the functionality of a car park structure or its components to an acceptable level.

**Robustness**   The ability of a structure subject to accidental or exceptional loading to sustain damage to some structural components without experiencing a disproportionate degree of overall distress or collapse.

**Routine Inspection**   Regularly scheduled visual inspection of a car park structure, cladding, edge protection and any other defined aspects of the car park.

**Routine maintenance**   A periodic activity intended to prevent or correct the effects of minor deterioration, degradation or mechanical wear of the structure or its components.

**Structural Appraisal**   Evaluation of the structural adequacy of a car park structure, cladding and/or edge protection taking into account its environment, likely usage, extent of deterioration and anticipated service life.

**Structural integrity**   The ability of structural components to act together as a single entity.

**Structural Investigation**   Examination of a car park structure or components usually involving materials testing and measurement.

[Part 1 of these Recommendations is intended for Owners and Operators]

# Part 1
# Car park structures, responsibilities, management and procurement

# 1. Introduction and scope

1.1 These Recommendations as a whole, i.e. those in both Part 1 and Part 2, relate to all types of multi-storey car park found in the UK for use by the public. They do not specifically consider car parks using mechanical stacking systems or private access car parks where different standards may be considered acceptable.

1.2 The recommended principles and approaches to management given in this Part 1 apply to car park structures generally. The more detailed technical recommendations in Part 2 relate mainly to concrete car park structures since they form a large majority of existing structures. To assist implementation of the approaches and techniques described, references are given to sources of more comprehensive technical, scientific and engineering information.

1.3 The Recommendations cover existing car park structures, i.e. primary structure, cladding and edge protection. They are concerned, in particular, with the safety of these structures whilst the car park is in service and with the actions needed by Owners, Operators and their professional engineering advisers to maintain safety and serviceability.

1.4 Lighting, lifts, security systems and other mechanical or electrical equipment and the safety of such equipment are not covered within the scope of these Recommendations. It may be appropriate and convenient to consider extension of the principles described in these Recommendations to such other systems and equipment.

1.5 The primary objective of Part 1 of these Recommendations is to enable the Owner or Operator on the basis of advice from the Engineer:

- to plan and implement an appropriate inspection and maintenance system;
- to know what to demand and expect of the Engineer and others involved in inspection and maintenance; and
- to understand and evaluate the proposals, reports and recommendations from them.

1.6 The Recommendations are issued by the National Steering Committee on Inspection of Multi-storey Car Parks of the Institution of Civil Engineers. The Committee was formed in 1998 following a number of recent structural failures and accidents and growing concern about deterioration in the condition of many car park structures.[1]

1.7 The Institution of Structural Engineers with the support of the Institution of Highways and Transportation, the Institution of Civil Engineers and the Chartered Institution of

Building Services Engineers has prepared the Third Edition of the report: 'Design recommendations for multi-storey and underground car parks'. It was published in June 2002 and includes up-to-date design recommendations and guidance on structural design, detailing and durability, and the design of barriers.[2] The aim of that guidance is to provide the basis for improved long-term performance of car park structures built in the future.

1.8   A study of design, specification and compliance testing of edge protection in multi-storey car parks carried out for the then Department of Environment, Transport and the Regions was completed in 2001.[3,4] In 2000, the then Department of Transport, Local Government and the Regions (DTLR) and a number of industry and university partners commenced work on a collaborative research project 'Enhancement of whole life performance of existing and future car parks'.[5] A main aim of that project was to analyse the performance of existing car park structures in order to assist the preparation of these Recommendations. The results of both of these projects are therefore incorporated, where appropriate, into these Recommendations.

1.9   It is recommended here that an Engineer be appointed to advise the Owner/Operator on the structural safety, inspection and maintenance of each existing car park structure. For this purpose, recommendations are given to assist the Engineer to use judgement in developing a long-term inspection and maintenance strategy for each multi-storey car park and in implementing it through a 'Life-care Plan'.

1.10   Throughout these Recommendations, where appropriate, recommended actions are addressed to the Owner and/or Operator. It is recognized that there are several possible contractual relationships between the Owner and Operator. The Recommendations should therefore be interpreted having regard to the relationship existing. Any new contract between Owner and Operator should be drafted to define clearly the allocation of responsibility for implementing the whole or individual parts of these Recommendations.

# 2. Car park structures

2.1 Local authorities own many multi-storey car parks. Other owners include property developers, private car park operators, airport operators and health trusts. Multi-storey car parks provide a service to the public and are often commercially important to the Owner/Operator through generation of revenue.

2.2 The majority of car park structures are freestanding independent structures, the internal space being devoted to accommodating parked vehicles and to their entry and egress. The car park may also be a part of a multi-use structure where other parts of the building are used for other purposes, e.g. retail premises, offices or residential accommodation.

2.3 Most multi-storey car parks in the UK are concrete column and deck structures (Figure 2.1). A considerable variety of structural types exists as a result of designing for faster and cheaper building and for fitting the structure into an awkward site or below a

*Figure 2.1   In situ reinforced concrete car park structure.*

building. The types existing are based on different structural concepts. The main types are:

- insitu reinforced concrete column and slab;
- insitu frame, precast prestressed deck;
- precast frame and deck;
- Lift-Slab construction.

2.4 Structures and decks below ground level are usually formed entirely in reinforced concrete.

2.5 There are also some older car park structures built with masonry load-bearing walls, concrete beams and voided deck slabs with cast-in pitch fibre pipes. In more recent times, new forms of structure have been built. There is a small number of car park structures built with bonded post-tensioned concrete decks. In several cases insitu beams post-tensioned using proprietary high tensile steel bars through precast external columns were used. Following developments in the technology, unbonded post-tensioned flat slabs have also been used in some recent construction.

2.6 A number of structural steel framed car parks has been constructed, partly as a result of relaxation of structural fire resistance requirements. The structural forms include steel frames with insitu or precast prestressed concrete decks and insitu joints and topping, metal composite decks and slim deck floors. A more detailed description of the main types of car park construction may be found elsewhere.[2]

2.7 A cladding, usually of concrete or masonry construction, may be present at each deck level. Barriers of metal or concrete construction are placed around the edges of decks to protect pedestrians and/or retain out-of-control vehicles within the car park. They may be integral with the cladding or, more usually, provided inboard of the cladding. The top parking level is usually without a roof.

2.8 The top parking deck is usually covered with a waterproofing layer but lower decks and ramps often have no waterproofing to the running surface. Decks and ramps of some older car park structures have been given surface treatments and/or waterproofing during refurbishment in recent years to reduce the risk of corrosion to embedded steel and to prolong serviceable life.

2.9 Car park structures have often received bad reports in the popular and technical press. The reasons include:

- The boom in multi-storey car park construction in the 1960s led to a spate of remedial actions in the 1970s and 1980s as structural distress was manifest by deflections, cracks, vibrations and spalling of reinforced concrete. These and other defects were attributed at the time to inadequate design and detailing, lack of adequate concrete cover to steel reinforcement due to poor design and construction, little provision of waterproofing to exposed deck surfaces, and inadequate maintenance often amounting to neglect.
- Car park structures are much more at risk from deterioration than normal buildings. In the 1960s, car park structures were treated in design and construction as buildings. They were designed to standards that are now known to be inadequate, particularly concerning provisions for durability. Durability of concrete structures was not well understood and de-icing salt was only then beginning to be widely used. The result has been that, for some car park structures, the life to first major maintenance has been much less than expected from standards used for design and

construction of buildings at the time. It is now well known that wetting and drying, freeze/thaw action and de-icing salt generally create conditions for rapid deterioration. Many car park structures were also poorly constructed, with poor drainage, maintenance and repair further aggravating deterioration. Consequently early action to improve durability through the use of waterproofing and coatings, and perhaps cathodic protection, can be a worthwhile investment.

- In addition to the costly early deterioration that occurred in some cases, several structural failures[1] and accidents[6] have occurred over recent years. The investigation of the partial collapse of the top floor of the Pipers Row car park Lift-Slab structure in 1997 revealed sensitivities to structural safety hitherto unrecognized.[7] Accidents involving failures of edge protection barriers to restrain vehicles have emphasized the need for better standards of barrier provision and maintenance.[1,6]

2.10    Car park structures are generally open to the weather, the large shaded deck areas are often damp, and vehicle-borne water contaminated with road de-icing salt and oil drips onto the horizontal surfaces (Figure 2.2). Although de-icing salt is only likely to be applied to ramps and top deck surfaces, experience has shown that significant amounts of salt are carried in and deposited on car park decks by vehicles when the surrounding highways have been salted. In coastal areas, car park structures are also subject to chloride exposure from sea water carried by the wind. They may also be subjected, regularly or occasionally, to liquids such as urine, antifreeze and liquids from broken bottles dropped by shoppers. The risk of corrosion of reinforcement in decks is increased in particular where the deck is sheltered from the rain and is not washed down. Drainage of contaminated surface water is often poor, especially in older structures. Thus car park structures, cladding and edge barriers are generally subjected to an aggressive environment. Their exposure is more akin to the severe conditions experienced by bridges than the protected environment of building structures.[2]

2.11    Some older car park structures also contain unusual and low strength details particularly sensitive to deterioration. De-icing salt is now less commonly used in car park structures but is inevitably still brought in on vehicles during cold weather when surrounding highways are salted. The potential effects on durability and structural

*Figure 2.2  Typical reinforced concrete car park structure.*

performance of commonly found defects in concrete car park structures are described further elsewhere.[5] Car park structures designed and built to the recently issued design recommendations[2] are likely to be more durable.

2.12 Cladding panels have not always proved to be durable and fixings have sometimes been found to be inadequate and prone to deterioration. Only in some cases were they designed for vehicle impact.

2.13 Edge protection barriers around the perimeter of car park decks have the following functions. They should protect pedestrians, especially small children, from accidentally falling over the edge; they should not be easily climbable by children. They should provide a reasonable protection against vehicles being driven, inadvertently or otherwise, over the edge. Barriers also serve in some locations to protect cladding, brickwork, stair towers, etc., from damage by vehicles.

2.14 Many early designs of edge protection barriers do not provide reliable containment. Some existing barriers are substandard and may also have deteriorating holding-down bolts and fixings. Sometimes replacement barriers have been installed incorrectly and not in accordance with the manufacturer's recommendations.

2.15 Background to the design, construction and performance of car park structures, cladding and edge barriers is described further in Appendix A.

7

# 3. Responsibilities of Owners and Operators

3.1 Multi-storey car parks should be safe for use by the public. They should also inspire public confidence and, to that end, they should provide a safe environment for the user (Figure 3.1). Many multi-storey car parks meet these requirements, but some do not.

3.2 The prime responsibilities for the safety of a car park structure, its cladding and edge protection lie with the Owner and Operator of the car park.

Figure 3.1 Large, recently constructed car park structure.

3.3 Guidance for businesses on corporate governance includes accounting for the actual value and future life costs on their building stock.[8]

3.4 The Owner/Operator should know the form of the structure. They should recognize that car park structures deteriorate over time, especially if not well maintained, and become unsafe in various ways unless appropriate actions are taken. The unprotected deck surfaces of most car parks are exposed to an environment that is more severe than is experienced by higher specification structures, such as highway bridges, which are waterproofed against the effects of de-icing salt brought onto the deck surface (Figure 3.2).

3.5 It is the responsibility of the Owner/Operator to arrange for the car park structure, cladding and edge protection to provide adequate standards of safety whilst in use. This objective can generally only be met through:

- Appointing an appropriately experienced Engineer to advise on inspection and maintenance, and to be the 'single responsible person' with overall responsibility for advice on the safety of the structure.
- A 'Life-care Plan' as advised by the Engineer, i.e. implementing a long-term strategy that includes Inspections, Condition Surveys, Structural Appraisals, maintenance and repair of the structure, focused on performance-critical areas where exposure is particularly severe, see Section 5.
- Arranging for continuity of engineering responsibility for safety with formal transfer of responsibility and all drawings and records of construction, inspection, appraisal and maintenance when changes in engineering personnel are made.
- Promptly referring to the Engineer for advice any reports of damage or structural distress that potentially may jeopardize safety.

Figure 3.2 Typical parking deck.

- Enabling all records and reports relating to the original design and construction of the car park and subsequent Initial Appraisals, Condition Surveys, Structural Appraisals, maintenance and repairs to be kept safely in an accessible location (see 5.4).

3.6 The actions needed for these purposes are described in Sections 5–8 below and indicated in Tables 5.1 and 6.1.

3.7 Organizations passing on ownership or responsibility for the operation of a multi-storey car park should provide all records relating to the design, construction, inspection, appraisal and maintenance of the structure, cladding and edge protection, and especially records of any Structural Appraisals, to their successor. Those taking on ownership or responsibility for operation of a multi-storey car park should try to obtain all such records.

3.8 To facilitate continuity of the Life-care Plan on change of ownership or responsibility for operation, any Owner or Operator engaging the services of an organization or individual (to provide services as an Engineer or Inspector or to carry out investigation services) should stipulate in the contract of engagement that any Structural Appraisal shall be deemed to be also for the benefit of any future Owner or Operator.

3.9 The allocation of responsibility between Owner and Operator for the safety of the structure and for implementing these Recommendations in whole or in part should be set down clearly in the contract between Owner and Operator. If existing contractual arrangements are unclear, the Owner should take steps to resolve the position.

# 4. Legal overview

4.1 Owners and Operators of multi-storey car parks have duties under law to provide and maintain their premises so that they do not pose risks to the safety of their employees, other persons using the premises as a place of work, visitors (including trespassers and children) and the public at large. Failure to discharge the various duties may give rise to criminal and/or civil liability, and may result in intervention by statutory authorities. The law in this area is complex and may be subject to change. Some relevant legislation is mentioned below.

4.2 The object of these Recommendations is to provide guidance on standards of care which constitute good practice and which should satisfy statutory and common law requirements. Advice should be sought on legal requirements that may be current at any particular time.

4.3 Car park structures should be built so that they fulfil the requirements for safety in the Building Regulations,[9] in particular in Part A: Structure; Part B: Fire; and Part M: Access for the Disabled.

4.4 There are also powers vested in Local Authorities under the Building Act 1984[10] and, in London, under the London Building Acts (Amendment) Act 1939: Part VII, Dangerous Structures, to intervene if the premises are in a defective state, if the building is in a dangerous condition or is used to carry such loads as to be dangerous. Similar powers are vested in Local Authorities in Scotland.

4.5 The Health and Safety at Work etc. Act 1974[11] imposes specific duties to maintain premises in a safe condition, under Sections 2 and 4. Section 2 imposes the duty on an employer to do whatever is reasonably practicable to maintain any place of work under his control so that it is safe and that the means of access to and egress from it are safe for his employees. Section 4 imposes a similar duty on any person who has to any extent control of non-domestic premises, to ensure the safety of persons who are not his employees but use the premises made available to them as a place of work. These duties are elaborated by the Workplace (Health, Safety and Welfare) Regulations 1992, particularly Regulation 5, Maintenance of Workplace etc., and Regulation 17, Organisation etc. of traffic routes.[12]

4.6 Section 3 of the Health and Safety at Work etc. Act 1974 imposes a broader duty on an employer (i.e. a person who has employees) 'to conduct his undertaking in such a way as to ensure, so far as reasonably practicable, that persons not in his employment are not thereby exposed to risk to their health and safety'. For an employer whose undertaking is the operation of a multi-storey car park, this could create duties to maintain the premises for the safety of visitors and the public who might be affected, e.g. passers by.

4.7 Breach of Sections 2, 3 and 4 of the Health and Safety at Work etc. Act 1974 or of the duties imposed by the Workplace (Health, Safety and Welfare) Regulations 1992 is a criminal offence. The Management of Health and Safety at Work Regulations 1999 are also relevant. It is for a person charged with an offence to prove that everything was done that was reasonably practicable. In addition, Health and Safety Executive Inspectors have powers to issue Improvement Notices and Prohibition Notices if they are of the opinion that there are breaches of the statutory requirements.

4.8 There is more specific occupational health and safety legislation that may be relevant, particularly, for example, during maintenance, repair, rehabilitation or replacement works. Regulations that may apply include the Construction (Design and Management) Regulations 1994 and the Construction (Health, Safety and Welfare) Regulations 1996.

4.9 Multi-storey car parks are premises subject to the Occupiers' Liability Acts 1957 and 1984.[13] Under the provisions of these Acts, the occupier of premises has a duty of care to arrange for the premises to be reasonably safe for people who enter them.

# 5. Life-care Plan

## 5.1 Management

5.1.1 The Owner and/or Operator is recommended to adopt a 'Life-care Plan' for each car park structure based on the advice of the Engineer. To that end, the Owner and/or Operator is strongly advised that an appropriate Engineer with experience of the deterioration of structures and knowledge of the materials science of deterioration processes is appointed to:

- take overall responsibility for advice on the structural safety of the car park structure, cladding and edge protection;
- determine what inspection, appraisal, maintenance and other works are required immediately or over time; and
- advise accordingly.

5.1.2 A rigorous inspection and maintenance process is essential to enable the safe and economic use of a car park structure. For this purpose, a continuous and progressive process with actions to take stock from time to time is recommended. The Engineer can provide it through the development and implementation of a Life-care Plan. Background to the concept of life care may be found elsewhere.[14] The main elements of a Life-care Plan for an existing car park are described below and outlined in Table 5.1.

## 5.2 Getting started

5.2.1 A continuous life-care process should be in place for each car park. A baseline for implementation is required. It is provided by an up-to-date Structural Appraisal undertaken by the Engineer. This should be based on an initial Condition Survey which can be used to benchmark the Structural Appraisals needed at intervals during the remaining life of the car park structure, cladding and edge protection.

5.2.2 The transition from the existing arrangements for inspection and maintenance to an appropriate Life-care Plan can be made along the following lines:

- For some car park structures, a recent Structural Appraisal (within the past 8 years) may serve as a baseline provided the Engineer is satisfied that it is sufficiently comprehensive and structural safety is unlikely to have been prejudiced in the meantime.

*Table 5.1. The main elements of a Life-care Plan for car park structures*

| Element | Work carried out by | Commissioned by and reporting to | Interval[a] |
|---|---|---|---|
| **Inspections** | | | |
| Daily Surveillance | On-site staff[b] | Car park manager | Daily |
| Routine Inspection | Inspector/assistant[b] | Car park manager | Less than 6 months[d] |
| Initial Appraisal and Condition Survey[e] | Engineer/Inspector[c] | Owner or Operator | Less than 8 years[d,e,f] |
| Structural Appraisal | Engineer[c] | Owner or Operator | Less than 16 years[d,e] |
| Special Inspection[g] | Engineer[c] | Owner or Operator | As required[g] |
| **Maintenance and repair**[f] | | | |
| Routine | On-site staff[b] | Car park manager | Monthly |
| Cosmetic/protective | Contractor | Owner | As advised by Engineer |
| Preventive/structural | Contractor | Owner | As advised by Engineer |
| **Rehabilitation and replacement** | Contractor | Owner | As advised by Engineer |

[a] The intervals used for any particular car park structure should be appropriate to the particular circumstances. They should be set on the advice of the Engineer.

[b] These staff are usually in the employ of the Operator, but they can be staff of an inspection consultant or contractor.

[c] The Engineer may be an employee of the Owner or Operator, or the work may be commissioned by the Owner/Operator from external consulting engineers or another suitably qualified and experienced body whose work is led by the Engineer and supported by an Inspector and assistants.

[d] Shorter intervals than the maximum values given are likely to be appropriate. The Engineer should advise the Owner/Operator taking into account the condition of the car park structure and the defects known to be present.

[e] Condition Survey may include Structural Investigation. The first Condition Survey of a new car park structure may normally be scheduled 3 years after completion and handover (see 5.2.2).

[f] For any car park structure over 3 years old where there are no adequate records of the structure, cladding and edge protection and/or of a Structural Appraisal in the last 8 years, a Structural Appraisal should be undertaken **without delay** (see 5.2.2).

[g] Special Inspection and immediate appraisal may be required in the event of a structural accident, fire, reported structural distress or suspected inadequacy (see 12.5).

- For others that are over 3 years old where there are no adequate records of the structure, cladding and edge protection and/or of a Structural Appraisal within the previous 8 years, then a Structural Appraisal should be undertaken **without delay**.
- For new structures (under 3 years old), the process can normally be started by a 'benchmark' Condition Survey at the age of 3 years leading to a Structural Appraisal as advised by the Engineer. During the construction liability period, it is important for Owners to identify defects that become apparent and could lead to future maintenance burdens.

## 5.3 The Plan

5.3.1 The use of a Life-care Plan for each car park structure enables implementation of a long-term inspection and maintenance strategy. The Owner/Operator is recommended to allocate adequate resources for the development and implementation of a Plan appropriate for each car park structure. It is especially important for sufficient resources to be provided for the first Initial Appraisal and Condition Survey. Otherwise, critical structural details and defects may be overlooked and an insufficient Condition Survey and Structural Appraisal made. Further actions may then be inadequate, resulting in inappropriate repair and substantially increased costs subsequently.

5.3.2    The Life-care Plan should be designed to keep the car park structure under inspection at appropriate intervals so that maintenance and repair interventions can be determined and made in the most timely and economic way and safety maintained. The Plan can be envisaged as a continuous progressive process of care. It defines the timing of inspection, appraisal, maintenance and repair works, and should be revised at intervals as recommended by the Engineer. It enables an integrated inspection, appraisal and maintenance programme. It allows the Owner/Operator to realize financial savings through the efficient management of resources, especially where a stock of car park structures is involved.

5.3.3    A collaborative effort is needed by the Owner and Operator, the Engineer and suppliers of inspection, maintenance and repair services to determine an optimum approach for the maintenance and safe operation of a multi-storey car park. To that end, the purpose of the Life-care Plan for the car park structures is to maintain structural safety in a cost-effective way and to avoid, if economically possible, a declining capital value due to deterioration and delayed maintenance. For the specific circumstances of each car park structure, the Plan should implement an optimum inspection and maintenance strategy developed by the Engineer based on Structural Appraisal.

5.3.4    Owners generally need to maintain continuity in the use of the car park so that the operations of connected facilities are not interrupted. To this end, the Life-care Plan is essential and, in today's competitive environment, may need to include rehabilitation to enable the car park to provide a secure and welcoming environment for the user.

5.3.5    Overall, the Life-care Plan advised by the Engineer should meet the requirements of the Owner and/or Operator and should be understood and used by them. In general, Owners and Operators will desire long-term serviceability of the structures achieved at minimum cost and interference with the ongoing use of the car park. Where Owners and Operators have a stock of multi-storey car parks, they may wish to have an overall Plan for the inspection and maintenance of the stock of structures. Prioritization of maintenance and repair, rehabilitation or replacement works may then be necessary in order to achieve optimal implementation within financial and other constraints, including consideration of residual-life costs. The Owner and Operator, where relevant, should discuss these factors with the Engineer and all other parties concerned to enable an appropriate Plan to be developed and applied. Published guidance relating to highway structures may be found useful.[15]

5.3.6    The implementation of the Life-care Plan, and especially its programming, will depend on the age and quality of the original design and construction, defects known to be present and whether previous Structural Appraisal is recent and sufficient. A progressive approach is needed, starting as indicated in 5.2, in which the available knowledge of the car park structure is built up to enable development and full implementation of the Plan. It will also need to be compatible with overall commercial plans for future use and maintenance of the car park as a whole.

5.3.7    The recommended elements of a Life-care Plan are Inspections (Daily Surveillance, Routine Inspections, Initial Appraisal and Condition Survey, Structural Appraisal) and maintenance, with repair, rehabilitation or replacement as appropriate. The elements and framework of a Plan are shown in Table 5.1. For each car park, the Life-care Plan should generally include:

(1) Inspections and Structural Appraisals:

- Daily Surveillance and reporting of damage or operational problems by on-site staff (see 11.2);
- Routine Inspections, usually by an assistant (see 11.3);
- periodic Initial Appraisal and Condition Survey at intervals by the Inspector or the Engineer (see Section 12). The Condition Survey may include a second stage, Structural Investigation, specified by the Engineer (see Section 13). In some cases, the Engineer may advise that the Condition Survey needs to be complemented by monitoring potentially sensitive parts of the structure at intervals over time in order to keep track and assess the changing condition of progressively deteriorating components in a cost-effective way;
- Structural Appraisals at intervals by an Engineer (see Section 14).

(2) Maintenance, protection, repair and rehabilitation or replacement:

- Routine maintenance, usually by on-site staff (see 15.2);
- protective and preventive repairs carried out through contracts for the application of cosmetic treatments, of protective treatments to slow deterioration and/or to control corrosion, and for structural repair, strengthening, rehabilitation or replacement (see 15.3);
- recording of works undertaken.

5.3.8 The Engineer should advise, for any particular car park structure, the timing of inspection actions and the intervals before each is repeated. Use of judgement is necessary bearing in mind the age of the structure and identified or suspected defects and deterioration, and taking account of practical constraints on the scheduling of maintenance and repair works and of commercial considerations. The intervals will usually need to be less than the maximum values indicated in Table 5.1. In general the longer intervals given in Table 5.1 will be appropriate only for relatively new structures. For many older structures shorter intervals will be judged necessary, especially where defects and/or progressive deterioration has been identified.

5.3.9 Prior to making decisions on inspection intervals and procedures, the Engineer should make an assessment of health and safety risks and structural safety risks (see 5.3.11). Where significant deterioration is identified, the timing of repair works should be related to estimates of deterioration rates, the structural safety implications, and criteria for minimum acceptable technical performance.

5.3.10 Some defects in concrete structures become apparent using straightforward visual inspection. However, more commonly, deterioration is only detected visually when the underlying deterioration has advanced significantly to become apparent at the surface. Often this is far too late to avoid major repairs. A variety of non-destructive testing (NDT) techniques is commonly used to 'look' beneath the surface for early warning signs (see Appendix C). Potential defects signalled by external visual signs can only be identified by an inspection based on knowledge of reinforcement detailing, structural behaviour and deterioration processes. The frequency and methodology of inspection should be tailored to suit the particular features of each car park structure.

5.3.11 Prior to inspection, maintenance, repair, rehabilitation and/or replacement works, the legal requirements for an appropriate and adequate assessment of health and safety risks to employees and others who may be affected by the work should be met.[16-18]

A structured approach to risk assessment is required, first identifying the hazards, then assessing the risks and finally determining steps to control the risks. Consideration should be given to possibilities for avoiding risks or to substituting lower risk processes. Safe systems of work should be defined and, where appropriate, protective equipment provided. Adequate supervision, training and information should be provided as part of risk control processes (see 16.1).

## 5.4 Records

5.4.1   The Owner/Operator is recommended to arrange for records of the history of each car park structure, cladding and edge protection to be kept and maintained up to date. Much time and money can be saved by Owners/Operators if adequate records are kept and made available throughout the life of each car park. The work of Inspectors and Engineers and the associated costs can be reduced substantially if reliable records can be consulted instead of gathering the required information by Condition Survey and Structural Investigation every time a Structural Appraisal is required.

5.4.2   The Owner/Operator is recommended therefore to arrange for:

- records to be kept of the original design and construction, of the inspection on completion prior to handover, and of maintenance, repairs and modifications carried out since construction;
- reports of Routine Inspections, Initial Appraisals, Condition Surveys, Structural Investigations and Structural Appraisals to be retained, including drawings prepared from surveys, photographs and test reports;
- as-built drawings to be used to record repairs and to be kept up to date as part of the Life-care Plan;
- the safe keeping of calculations;
- all this information to be referenced in the Health and Safety file for the car park.

5.4.3   It is advantageous for records to be the responsibility of a designated keeper and filed in a computerized asset management database.

5.4.4   A free flow of record information between the Owner/Operator and the Engineer is essential for efficient implementation of a Life-care Plan. It is recommended that relevant records should be made available on commissioning the Engineer/Inspector to undertake the Initial Appraisal, Condition Survey or Structural Appraisal. Their access to records will assist development and implementation of the Life-care Plan. Where copies are not provided, increased investigation costs and disruption on site are likely to arise.

5.4.5   Consideration should be given to including a requirement in commissions to Inspectors and Engineers for reports to be produced in electronic as well as hard copy form. New documents produced as the Life-care Plan proceeds should be added to the records. Often records do not exist in sufficient detail, e.g. as-built drawings are not available, to provide the basis for Structural Appraisal. In these cases, the deficiencies in information on the relevant parts of the car park structure need to be made good during Condition Survey using, where necessary, Structural Investigation (see Sections 12 and 13).

# 6. Procurement of Initial Appraisals, Condition Surveys, Structural Investigations and Structural Appraisals

6.1 The Engineer's role in advising the Owner/Operator is to highlight any defects in the structure, cladding or barriers, to advise what inspection is required, to arrange for necessary survey and investigation, to appraise, and to recommend to the Owner or Operator options for repair and other actions necessary for safety and continued use of the car park in the short and long term. An Inspector working with the Engineer or working for external consulting engineers or other suitable body may undertake some of the work.

6.2 Owners and Operators, in selecting the Engineer, Inspector, and other competent persons and organizations, should bear in mind the need to keep communication difficulties at a minimum. The disadvantages and risks of uncoordinated contracting of different tasks of inspection, investigation and appraisal from separate organizations should be minimized. For example, using organizations that are competent to undertake several of the required tasks can reduce the risks of ineffective communication. Similarly, Owners and Operators are recommended to arrange effective information transfer and briefing when tasks are taken over by new personnel or organizations. Provisions for coordination should also be made where separate organizations are working on different but related aspects of a Life-care Plan.

6.3 Overall, a wide range of knowledge and experience is needed for implementation of a Life-care Plan. Those appointed should be selected so that the requisite skills are available in the team. Key actions are summarized in Table 6.1. The structural inadequacies and deterioration to which many car park structures are prone usually require the application of engineering knowledge of early design approaches that have now been superseded and of specialist materials and inspection expertise. For this purpose, specialist knowledge of deterioration processes, the materials science involved, modern techniques for reducing rates of deterioration and for making effective repairs is needed. A broad knowledge of maintenance, repair and rehabilitation options is also

*Table 6.1   Typical sequence of recommended key actions for inspection, maintenance and management of a car park structure*

**Owner/Operator**

(1)   For each car park structure, its cladding and edge protection appoint an Engineer to advise and to act as a single 'responsible person' for structural safety and the development and implementation of the Life-care Plan (see 5.1 and 5.2).

(2)   Where required, arrange appointment of an Inspector (see 5.3).

(3)   Make arrangements for records to be kept by a designated 'keeper' (see 5.4).

(4)   Provide the Inspector/Engineer with all records relating to the design and construction of the structure (including as-built drawings, calculations – if available – materials certificates and test results, etc.) and reports of previous Initial Appraisals, Condition Surveys, Structural Investigations and Appraisals (see 5.4).

(5)   Arrange provision of suitable checklist systems for Daily Surveillance by on-site staff and Routine Inspections by an assistant, the Inspector or the Engineer (see 11.2 and 11.3).

**Inspector**

(6)   Carry out Routine Inspections at specified intervals, or supervise this work being undertaken by an assistant (alternatively, depending on organizational arrangements, supervision may be by the Engineer), using the checklist prepared and agreed with the Engineer (see 11.3).

(7)   Carry out Initial Appraisal: assess sufficiency of records and carry out any additional site measurement, inspection etc. deemed necessary. In discussion with the Engineer, identify all locations and details that are critical to structural integrity (see 12.2).

(8)   Carry out Condition Survey, i.e. a full visual inspection and sampling and testing, e.g. half-cell, cover and chloride profiles, of the structure, recording deterioration with specific reference to the implications – deterioration affecting structural safety and condition to be separately recorded (see 12.3).

(9)   In consultation with the Engineer, extend the Condition Survey by Structural Investigation to provide any further information needed to enable Structural Appraisal to be completed (see 13.1).

**Engineer**

(10)   Review Initial Appraisal and Condition Survey and advise on next steps (see 12.4).

(11)   Undertake Structural Appraisal giving evaluation of integrity and adequacy of structure, cladding and edge protection. Produce summary of the results with assessed structural capacities being compared with those required at all critical locations (see Section 14).

(12)   Estimate residual life of structure, cladding and barriers and appraise the implications of further deterioration. Report to Owner/Operator (see Sections 14 and 17).

(13)   If the structure is deficient in strength, provide costed options for interim measures, pending repair works, together with costed options for repair works as may be requested by the Owner/Operator (see Section 14).

(14)   Produce costed options for long-term inspection, maintenance and periodic reappraisal of the structure, cladding and barriers, following remedial works, with specific reference to ensuring continued structural safety and reducing rates of deterioration (see Section 14). Update Life-care Plan. Report to Owner/Operator.

**Owner/Operator**

(15)   Review reports from Inspector and Engineer and consider options (see Section 7). Select option in discussion with the Engineer.

(16)   Commission maintenance and repair/rehabilitation works, if needed. Where repair proposals have been prepared by a contractor, ask the Engineer to review them prior to commissioning (see Section 8).

(17)   Consider proposed update of Life-care Plan and commission continuing actions on the basis of the update accepted (see 14.4 and 14.5).

needed, together with an ability to provide realistic cost estimates. In Condition Surveys and Structural Investigations it is recommended that the Engineer should work closely with materials and repair specialists.

6.4   The expertise brought to bear should be appropriate to the circumstances in each car park. Appropriate expertise may be found particularly, but not exclusively, amongst professional engineers and materials and repair specialists who have experience of advising clients on the maintenance of car park structures or concrete bridges. Appropriate specialists may be identified through professional, trade and industry bodies (see Section 20).

6.5    It is particularly important to allocate adequate resources for the Initial Appraisal, Condition Survey and Structural Investigation of defects and deterioration. Resource requirements for the Investigation can only be estimated very approximately prior to the Condition Survey.

6.6    Correct identification of a car park structure and its condition is essential for the implementation of reliable remedial works. In particular it is crucial to identify deck structures that may be prone to sudden failure at column supports or to removal of a column by vehicle impact and potentially then to be vulnerable to progressive collapse (see Section 14). Owners and Operators should be aware that some forms of reinforcement corrosion produce no visible signs at the surface of the concrete although the structure may be significantly weakened (see B2). Without correct identification of dimensions and structural condition (Figure 6.1), ineffective repairs may be made that do not restore structural safety and have to be remade with consequent additional expense and disruption. Inadequate or inappropriate Investigation of deterioration may also result in the unanticipated need to extend remedial works whilst they are in progress.

*Figure 6.1    On-site measurement of half-cell potentials.*

6.7 Where concrete slab structures have been repaired previously without the involvement of an Engineer, propping has often been substandard. Ineffective propping can cause a redistribution of moment that can lead to overstress and excessive deflection/cracking. Owners/Operators should not rely upon superficial concrete repairs or propping without advice from the Engineer beforehand.

6.8 The Owner or Operator commissioning the Engineer and other competent persons should properly consider and review their proposals, reports and recommendations and act upon them.

# 7. Options for maintenance, repair, rehabilitation and replacement

## 7.1 Assessing options for concrete structures

7.1.1 The selection of the most appropriate option will depend mainly on the Owner's/ Operator's business objectives, the requirement for public safety and the results of Structural Appraisal.

7.1.2 Advice from the Engineer following Structural Appraisal should identify and assess the options. Important factors for a concrete car park structure are likely to include the amount of chloride contamination and carbonation, the extent of deterioration and/or inadequate structural strength, susceptibility to progressive collapse, accessibility and aesthetic requirements.

7.1.3 A large number of options for maintenance and repair are potentially available but few will be appropriate for the Owner/Operator in any particular case. The options may range from doing nothing now to replacement of part or the whole of the car park structure, cladding and/or edge barriers. The many options between these two extremes may include undertaking repair when an agreed level of deterioration is identified through monitoring its progress, applying preventive measures to prolong structure life by delaying the onset of deterioration or by slowing its progress, e.g. through improved routine maintenance, cosmetic repair, continuous preventive maintenance or protective treatments.

7.1.4 The options considered will depend on the outcome of the Appraisal, in particular any structural inadequacies or sensitivities identified and the predictions of trends in deterioration of the materials and structure. Candidate options for the particular circumstances will need to be identified and evaluated (see Section 14). The overall strategy of the Owner/Operator will attract particular options. For concrete structures, approaches are given in BS EN 1504: Part 9 and elsewhere.[19–21]

7.1.5 Improved routine maintenance may be needed where, for example, there is persistent ponding and blockage of drains. Monitoring may be needed to check rates of structural deterioration or damage.[22–27] Continuous preventive maintenance and protective treatments, such as the application of waterproofing treatments to exposed concrete deck and ramp surfaces, may be a cost-effective way of slowing down deterioration processes and thus extending life before repair. Repair is generally necessary where reinforcement corrosion is apparent and concrete spalling is occurring (Figure 7.1). In some cases it may be necessary for this to be carried out as soon as possible. In other cases it will be possible to delay the work until a more convenient time. Alternatively, where corrosion is advanced widely in the structure, it may be decided to remove loose concrete only and to allow degradation of the structure, with temporary propping where necessary, until it is replaced.

7.1.6    In considering options for maintenance, repair and rehabilitation or replacement, it is generally advantageous first to take a strategic long-term view and to evaluate options in overall residual-life cost terms (see 7.2). Costs of candidate options needed to keep the car park structure in a serviceable condition are generally the most important consideration in deciding on the option to be adopted.

7.1.7    Where a specialist contractor or materials supplier defines repair needs, it is recommended that the Engineer be commissioned to undertake an independent review of the repair proposals.

*Figure 7.1    Soffit of deck slab after removal of cracked concrete and preparation for the application of patch repair material.*

## 7.2 Residual life assessment of options

7.2.1 Traditionally maintenance and repair work on structures has been undertaken following inspections when the structure is found to be near to or approaching an unacceptable condition, or to be below the acceptable safety level. However, this approach does not include consideration of preventive action before critical levels of safety or deterioration are reached. Preventive action is often a part of current practice and the Engineer can provide an evaluation of what is worthwhile. Factors that may need to be considered are discussed in Section 15.

7.2.2 Minimizing residual life costs whilst maintaining safety may be an appropriate aim for an Owner/Operator. This approach enables sufficient and justifiable amounts of preventive work to be identified and carried out over the years so that financial burdens of rehabilitation and replacement are not unduly postponed. Guidance on procedures for identifying options based on whole life costs as far as practicable is being developed for highway bridges and structures.[15] Similar procedures may be adopted for deciding on options for car park structures. Sustainability may also need to be considered.

# 8. Specifications and contracts for Condition Survey, Structural Appraisal, maintenance and repair works

## 8.1 Condition Survey and Structural Appraisal

8.1.1 The scope of the required Survey, including any Investigation, and Appraisal should be clearly defined.[28] It is often best to proceed in stages, each stage being defined on the basis of what has been found in the previous stage. The work may be commissioned by the Engineer or by the Owner/Operator on the recommendation of the Engineer.

8.1.2 The work required will depend first on the Initial Appraisal. The provision of access should be considered early in the planning. In some cases, a materials testing regime may be needed to identify defects, degradation, build-up of de-icing salt and the corrosion condition of the reinforcement to assist in the design of appropriate repair works. The diagnosis of condition and rate of degradation may be assisted by monitoring condition at intervals over a period of time.[22–27]

## 8.2 Maintenance and repair works

8.2.1 The scope of the proposed works should be clearly defined. They should follow recommendations of good practice as given, for example, elsewhere.[16–18]

8.2.2 All proposals for maintenance, repair, rehabilitation or replacement should be subject to the advice of the Engineer. If the work required is to be tendered, it is recommended that the Engineer (responsible for formulating the scope of the work approved by the Owner or Operator) should prepare and clearly define in appropriate specifications and drawings, the works required. The Engineer should also produce designers' risk assessments, and have a continued supervisory role during execution of the work. This should include vetting any alternative proprietary products that may be proposed. The works may comprise one or more of the following parts of the Life-care Plan:

- preventive maintenance and cosmetic/protective repairs;
- preventive and structural repair works.

8.2.3 Maintenance and repair works should be entrusted only to contractors recognized as competent with appropriate skills and experience, under the direction of a suitably experienced engineer, and with adequate resources to complete the works as required. Repairs that may temporarily reduce structural strength, i.e. structural repairs, should be specified and made so that structural safety is not impaired.

8.2.4 If a specialist contractor or materials supplier has formulated the scope of the work, it is recommended that the Owner/Operator should seek an independent review by the Engineer before proceeding.

# 9. Main Recommendations of Part I

9.1 Owners/Operators are recommended to:

(1) appoint an appropriately experienced Engineer to advise on structural safety, inspection, maintenance and repair in accordance with industry recommended practice;

(2) request the Engineer to establish a Life-care Plan for each car park structure and the baseline for commencing implementation;

(3) provide adequate resources for implementing the Life-care Plan for maintaining the car park structure in a safe condition fit for use by the public;

(4) arrange for records of the design, construction, inspections, maintenance, repair, rehabilitation and replacement to be kept, maintained up to date, and made available to the Engineer and others as appropriate, as the life-care process progresses.

# Part 2
# Structural defects and deterioration, and implementation of Life-care Plan

[Part 2 of the Recommendations is intended for the professional engineering advisers to Owners and Operators.]

# 10. Structural defects and deterioration

10.1 Many early car park structures, i.e. those built in the 1960s and earlier, have required remedial action after only a few years in service. When they were built there was no perception of the need to specify higher standards for structural design and other requirements for car park structures than were used in current practice for buildings. The 60-year expected life to first major maintenance of buildings was not generally achieved. As experience has been gained over the years, it has become evident that higher standards are required and that car park structures should not be treated as ordinary buildings. Consequently, design standards and other requirements have been enhanced over time[2] (see Appendix A).

10.2 Generally it is early concrete structures that are now in the most deteriorated condition. This is due to their longer exposure to the aggressive environment, and due to poor standards of design and construction relative to modern standards and requirements. Structural deterioration was often manifest during the first years of the car park's life by excessive deflections, cracks (e.g. drying shrinkage and plastic settlement cracking) and/or spalling of concrete. These and structural defects were generally attributed to inadequate design and detailing, lack of adequate concrete cover to steel reinforcement due to poor design and/or construction, little provision of waterproofing to exposed surfaces and inadequate maintenance. At the same time early car parks have often been found to have structural design safety shortcomings in their structures, claddings or edge protection systems.

10.3 Robustness of car park structures was often not considered explicitly in early designs. Early structures may therefore be susceptible to progressive collapse, where there is deterioration or repair in the region of column/slab supports and/or the connections between structural components have not been well designed and made. Some concrete column/slab structures may be susceptible to failure suddenly by punching shear at slab supports, in particular after weakening by materials deterioration, e.g. pitting corrosion of reinforcement, by inappropriate repair or by ineffective propping during repairs.[7]

10.4 Edge barriers have often not been fitted as part of the main construction contract and were fitted subsequently. In some cases they were fitted without considering how vehicle impact loads would be transferred back into the structural frame and with little concept of problems that could occur when reinforcement is damaged by drilling for fixings or corrosion occurs. Post-construction drilling to install fixings increases susceptibility to water and chemical ingress, with consequent accelerated deterioration.

10.5 The standards of design and specification of car park structures have improved to some extent since the 1960s as BS 8110 was adopted[7] and understanding of the deterioration potential of de-icing salts and freeze/thaw action on concrete structures and of the severity of the environmental conditions in car parks has grown. Premature deterioration and structural design inadequacies have nevertheless been reported in some car park structures of recent construction. Within the national stock of multi-storey car parks as a whole therefore, structural safety inadequacies and deterioration may be found, the shortcomings tending to be more extensive in older structures. Past, and to a lesser extent, present standards of design and specification have not given adequate guidance on avoiding and minimizing such shortcomings.[7] The more stringent design requirements, e.g. enhanced lateral bracing and improved punching shear resistance, introduced as codes have been improved, could not have been foreseen in early car park structure design. Deficiencies that may be found in existing structures include:

- Penetration of water contaminated by de-icing salt into concrete causing corrosion of embedded steel reinforcement and – often but not always – cracking, spalling and delamination of the concrete cover. These forms of deterioration are often associated with:
  (a) a lack of appreciation of the risks, resulting in poor detailing and a low quality of the original construction (e.g. inadequate concrete quality);
  (b) low concrete cover to reinforcement;
  (c) plastic settlement of concrete after casting;
  (d) restrained drying shrinkage causing cracking of the concrete deck particularly where the slab is restrained at each end by a lift/stair core; and
  (e) leakage through slabs and joints arising from poor drainage of parking decks, incorrect joint installation and/or damage to joints during installation of services, e.g. electrical conduit.
- Freeze/thaw damage to concrete, particularly in top decks that are frequently saturated leading to surface scaling and reduced slab strength.
- Tendons in incompletely grouted ducts in post-tensioned concrete beams vulnerable to corrosion as a result of inadequate protection.[29]
- Structural concrete elements cast with calcium chloride additive in the concrete.
- Sub-standard fixing of reinforcement.
- Concrete cracking and corrosion of reinforcement due to carbonation of concrete.
- Concrete elements affected by alkali silica reaction.
- Inadequate punching shear capacity of slabs at columns in some structures built to codes pre BS 8110: Part 1: 1985 (see 14.4).
- Precast beams inadequately supported by deteriorating corbels.
- Reduction of strength due to inadequate/deteriorating concrete repairs.
- Primary structures built without specific consideration of key elements in the design.
- Inadequate fixings of cladding.
- Inadequate strength of edge barriers to restrain vehicles from falling over the edge following accidental impact.
- Inadequate pedestrian guarding where installed prior to BS 6399: Part 1: 1986.

10.6 It may not be safe or cost-effective to wait until significant visible deterioration of a concrete car park structure is present before taking remedial action. Simple repairs made to areas of visible damage in general are not likely to slow down deterioration and provide a significant extension to useful life – incipient anode effects often cause subsequent corrosion around the repaired areas.[7] Such repairs are often inappropriate and/or incorrectly applied. Structural Appraisal by the Engineer as part of a Life-care

Plan is needed first to diagnose the causes and assess the extent of structural inadequacy and deterioration and needs for repair. Appropriate repairs can then extend the serviceable life of the structure substantially and maintain safety cost-effectively.

10.7 Many forms of structural defect and deterioration of materials in concrete, steel or masonry structures of multi-storey car parks become visible to the naked eye before the structure becomes unsafe. However, this is not always the case. Whilst car park structures are more open to visual inspection than buildings, there are some situations in which safety can be placed in jeopardy without any visible manifestation. The deteriorating component may be hidden from view. Components may be hidden behind cladding or coatings, or encased in concrete, e.g. fixings behind cladding. In addition, although not common, severe local pitting corrosion of steel reinforcement in concrete decks can proceed unnoticed where heavy chloride ion penetration of the concrete is present. In the particular case of post-tensioned construction, severe corrosion of steel tendons may occur without external visible signs. These possibilities should be borne in mind when car park structures are inspected or appraised.

10.8 Some early multi-storey car parks included a parapet around the edges of the decks as part of the cladding to the structure. The parapet was assumed to be sufficient to restrain vehicles. In other cases the cladding was primarily decorative and edge barriers for vehicle restraint were only a concrete 'bump' stop near to the deck edge with a lightweight metal (and sometimes timber) balustrade for pedestrian edge protection. Sometimes barriers were fixed under separate contract after the main structure had been completed. As a result, reinforcement in the concrete was sometimes damaged during the installation of holding-down bolts. The design of deck edge protection has gradually improved over time as standards have been established (see Appendix A), but deterioration, especially of fixings, continues as a factor diminishing restraint capacity over time.

10.9 The structural defects, inadequacies and materials deterioration that may be present in car park structures, cladding and edge barriers are described more fully in Appendices A and B. Defects and inadequacies that are frequently found in concrete car park structures are described in more detail elsewhere.[5]

10.10 It is the task of Initial Appraisal, Condition Survey and Structural Investigation, as outlined in Sections 12 and 13, to identify deficiencies, especially potentially vulnerable components and features that may be prejudicial to safety. It is the task of Structural Appraisal, described in Section 14, to diagnose causes and implications for safety and future use of the car park.

10.11 Structural inadequacies, if present, should be identified and the causes and extent of deterioration diagnosed in Structural Appraisal. This information provides the basis for specifying appropriate repairs. Works can range from the application of a protective coating over minor deterioration damage to repair of parts of components or replacement of whole components, e.g. slabs, cladding panels or edge barriers. Potential deterioration of structure, cladding and edge barriers will become apparent in Condition Surveys once a car park is a few years old. Trends in chloride ingress, cracking, freeze/thaw damage, water seepage, carbonation and traffic wear can be estimated by Structural Investigation using sampling in selected areas. Future patterns of deterioration can be postulated[20] to assist with Life-care Planning of future inspections, maintenance and long-term budgeting. This work may enable pre-emptive action to be identified and priority to be given to reducing the rate of deterioration at an early stage. Decisions on life-care options should be made after careful consideration of their cost effectiveness and likely long-term performance.

# 11. Daily Surveillance and Routine Inspection

## 11.1 Preparation

11.1.1 Suitable checklist systems, either paper or computer based, should be provided for making reports of structure damage, equipment breakdown, or incidents observed during Daily Surveillance and Routine Inspections. The checklists should be prepared in consultation with the Engineer.

## 11.2 Daily surveillance

11.2.1 The Operator's on-site staff should be required to keep the car park and its equipment under Surveillance on a daily basis and to report any breakdown of equipment, obvious damage, e.g. to edge barriers, and any other untoward incidents in the use of the car park. The reports should be reviewed by the manager of the car park and reported to the Inspector or Engineer where significant damage appears to be present. It may be necessary to carry out a risk assessment to inform subsequent decisions on repair or other action required. Daily Surveillance is the first element of a Life-care Plan for the continuing safe operation of a car park. The Engineer is recommended to advise that a schedule of reported damage be kept.

## 11.3 Routine inspection

11.3.1 The next element of the Life-care Plan should be Routine Inspections, usually by an assistant working under the supervision of the Engineer or the Inspector. Routine Inspections are visual and should cover the structural frame, cladding and edge protection. They may also cover other defined aspects of multi-storey car parks that are outside the scope of these Recommendations, e.g. security, lighting and traffic management (see 1.3).

11.3.2 Routine Inspections should be scheduled regularly at less than 6-monthly intervals (see Table 5.1). Short intervals may be judged necessary by the Engineer, for example, if there is evidence of potentially significant deterioration in key load-bearing areas. These visual inspections should be based on a checklist (usually prepared by the Inspector) in order to identify all relevant items for inspection. The checklist should be developed on the basis of an assessment of safety risks determined in an Initial Appraisal (and Condition Survey if one has been carried out) (see 12.2 and 12.3). Defects highlighted by the checklist should be reviewed by the Inspector or Engineer and reported to the Owner/Operator where appropriate.

11.3.3 The recommended items of Routine Inspection of the structures of a car park include:

(1) Parking decks and ramps:
- check drainage, especially bottom ramp drains and gully outlets,[5] and report any blocked drains and/or ponding on decks or leakage through decks and consequential lime staining;
- check beams, columns and deck soffits for rust staining, damage, cracking or spalling;
- check trafficked surfaces and report any damage observed.

(2) Cladding:
- visually check and report any apparent movement, damage due to deterioration or vandalism, spalling, evidence of corrosion or dislodged elements. Inspect fixings where possible.

(3) Edge protection barriers:[5]
- check for damage due to accidental vehicle impact, vandalism or deterioration;
- check holding-down bolts and report any missing, ponding and/or signs of deterioration.

(4) Enclosures such as staircases, lobby areas, lift enclosures and plant rooms:
- visually check for signs of deterioration or damage due to vandalism or vehicle impact.

11.3.4 In cases where structural refurbishment has already been carried out, Routine Inspection may include the taking of observations from installed monitoring equipment.

11.3.5 The Engineer is recommended to advise the Owner/Operator to keep copies of reports of Routine Inspections for future review.

# 12. Initial Appraisal, Condition Survey and Special Inspection

## 12.1 General

12.1.1    To provide a basis for maintaining safety and for prolonging the life of a car park structure in the most cost-effective way, an Initial Appraisal of the structure, cladding and edge barriers followed by a Condition Survey should be carried out periodically by an Inspector or the Engineer (see Table 5.1). The Initial Appraisal and Condition Survey provide essential background for Structural Appraisal as a part of development and implementation of the Life-care Plan. Together with Structural Appraisal, they form the basis for determining options for future inspection, routine maintenance, repairs, strengthening, and rehabilitation or replacement (Figure 12.1).

## 12.2 Initial Appraisal of the structure, cladding and edge protection

12.2.1    The Initial Appraisal is essentially an overall structural and materials review by a desk study of records prior to a Condition Survey. It should include assessment of safety risks to personnel on site carrying out a subsequent Condition Survey, and to members of the public (see Section 16). Prior to the Condition Survey, the Inspector may need to establish the site conditions and to decide whether finishes have to be removed to expose the structure, e.g. deck surfacing. A contractor may need to be employed to provide means of access, e.g. for inspection of cladding, to expose the structure and to make good after the Survey (Figure 12.2).

12.2.2    The desk study should review the records provided by the Owner/Operator of the original design and construction, of the inspection on completion prior to handover, and of the maintenance and repairs carried out since construction. Reports of earlier Structural Appraisals and recent Routine Inspections should also be examined. As-built drawings, if available, should also be examined and copies prepared for updating with modifications, barrier changes or service penetrations made since the structure was first built. In the absence of as-built drawings and calculations, there will be a need for Structural Investigation at potentially critical locations. Removal of concrete cover to expose reinforcement and the use of suitable scanning techniques may be necessary

35

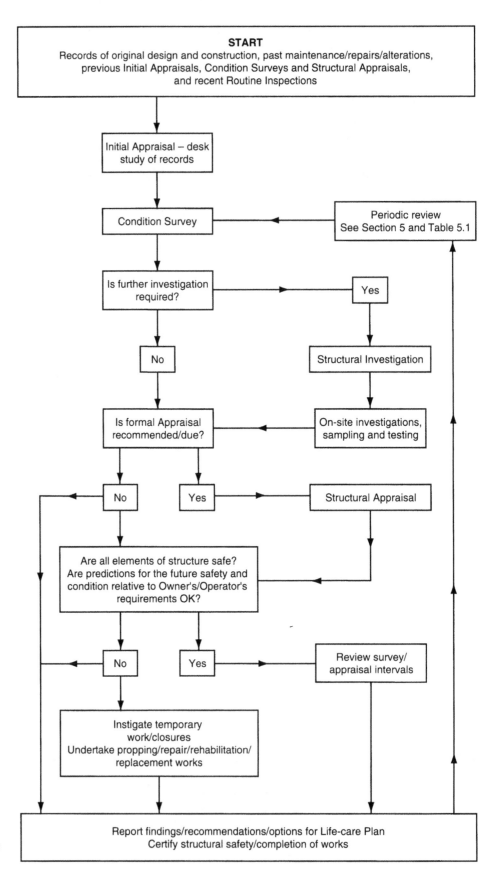

*Figure 12.1 The process of periodic Condition Survey and Structural Appraisal.*

Figure 12.2 Unusual cladding on a modern car park structure.

to establish the amount of reinforcement present (see Appendix C), so that 'back-analysis' of the structure can be carried out. The adequacy of new or replacement edge barriers and their fixings should be checked.

12.2.3    The Initial Appraisal will enable the Engineer, or the Inspector in discussion with the Engineer, to identify likely structural inadequacies and sensitivities. Locations in the structure that are likely to be most prone to deterioration can also be pinpointed, and the possible adverse implications on structural sensitivity and safety can be defined. Locations that are critical to structural integrity should be identified by review of the form of construction. Some locations may be difficult to inspect and need to be identified so that suitable inspection techniques can be called up in the Condition Survey. This approach enables the Survey, and also Routine Inspections, to focus in the most relevant and efficient way on areas that are structurally sensitive and/or appear to have deteriorated most. Features that should be identified in the Initial Appraisal include:

- the form of the structure, its cladding and barriers, including fixings;
- areas where conditions of local exposure are likely to concentrate deterioration due to water ponding and seepage, chlorides, urea, freeze/thaw action, carbonation and traffic abrasion;
- water leakage through slabs and construction or movement joints;
- cracking, spalling or signs of reinforcement corrosion in deck slabs, ramps, columns, corbels and cladding;
- structural details critical to the integrity of the structure and also those that are sensitive to poor construction and/or deterioration;
- details that may be structurally inadequate owing to limitation in the original design;
- signs of movement or excessive deflection;
- difficult to inspect components and areas needing specific survey or testing.

12.2.4    The Initial Appraisal is vital to the identification, by the Condition Survey, of the causes and extent of deterioration and defects, especially active corrosion of reinforcement or freeze/thaw action at structurally sensitive details. A checklist should be prepared for use in the Condition Survey.

## 12.3  Condition Survey

12.3.1    The Condition Survey should be based on the Initial Appraisal of the structure, cladding and barriers and, in the first instance, is effectively a 'benchmark' survey.

12.3.2    Condition Surveys are usually visual in the first instance and based on a checklist developed during the desk study (Figure 12.3). Where the Survey is extended to include Structural Investigation, the scope of the required Investigation will need to be specified. A materials and structural testing specialist should be employed to undertake the work. To establish a benchmark condition, a Condition Survey should include non-destruction testing (NDT) and analysis of samples, e.g. cover, half-cell and chloride profiles (see 5.2).

12.3.3    The Condition Survey should include photographs and measurement of visible evidence of defects and deterioration that affect structural safety or condition. The risk of omissions in the work may be reduced by use of the checklist. The following should be recorded:

- signs of movement of structural components, especially in the vicinity of sensitive structural details and joints, cladding elements and edge barriers;
- the results of close scrutiny of locations that are critical to structural integrity, especially where structural vulnerability may be present with aggravated deterioration. Where these locations are not visible, Structural Investigation may be needed;

*Figure 12.3  Identification of deck cracking during Condition Survey.*

- rust staining and cracks in concrete;
- spalling of concrete;
- hollow or delaminated concrete surfaces;
- evidence of corrosion of reinforcing steel (structure, cladding and edge barriers);
- areas of previously completed repairs and the areas immediately surrounding them;
- for steel-framed structures: impact damage, prising of connections due to corrosion, water penetration at joints and condition of concrete/steelwork interfaces;
- contamination on decks and other surfaces;
- wet or damp areas, especially ponding on decks and leakage, e.g. through expansion joints and from drainage systems;
- edge protection: corrosion of holding-down bolts and fixings, broken, lost or loose connections in barriers, and impact damage.

12.3.4   The records of the Survey should then be used to update relevant drawings and record deterioration and previous repairs. The Survey should provide any additional measurement needed for the subsequent Structural Appraisal of adequacy. For Appraisal, as-built reinforcement drawings of sensitive and deteriorated areas of the car park structure, marked up with modifications and details of previous repairs, are required.

12.3.5   Several test techniques are available for Structural Investigation, in particular for concrete structures and components, that can be used to extend a visual Condition Survey (see Section 13). These techniques enable more information to be collected on the extent and causes of deterioration damage and on the future risk of reinforcement corrosion or other degradation.

12.3.6   When making a Survey of deteriorating concrete components, screening into categories (visible deterioration/no visible deterioration, etc.) can provide a useful preliminary way of focusing towards areas needing closer investigation.

12.3.7   More detailed information on commonly occurring defects and deterioration in car park structures, cladding and barriers that may be present and impair structural performance and durability is given elsewhere.[5]

12.3.8   The results of the Condition Survey should be reported to the Owner/Operator (see 17.2).

12.3.9   The interval before the next Condition Surveys should normally be decided by consultation between the Owner/Operator, the Engineer and the Inspector after a Structural Appraisal has been completed. Account should be taken of the age of the structure, its design, and the defects and inadequacies identified in the primary structure, its cladding and edge barriers. The timing should be based on the framework given in Table 5.1.

## 12.4  Summary of principal actions in a typical Initial Appraisal and Condition Survey

12.4.1   The Condition Survey following Initial Appraisal is often in two stages, the second being Structural Investigation used to extend the information gained in the first visual stage of the Survey. Actions should be taken as required by the Engineer. They will normally include:

- Studies of the records of design, construction and maintenance of the structure, cladding, edge barriers and fixings. Assessment of the sufficiency of the information and making any additional site measurement, inspection, etc., deemed necessary. Where as-built drawings are not available, the relevant parts of the car park structure should be surveyed and drawings produced to record positions and extent of cracks and defects.
- Appraisal of the structure on the basis of the records and identification of all locations, details, etc., that are, or may be, critical to structural integrity, including those where deterioration of materials may exacerbate criticality.
- Preparation of a checklist for the Condition Survey.
- A visual inspection of the structure, cladding, edge barriers and fixings recording visual deterioration and damage caused by impact, with specific reference to the implications.
- Making any Structural Investigation necessary to provide further information needed to enable a Structural Appraisal to be completed.
- For a structure with a history of deterioration problems, monitoring of the progress of deterioration through regular testing and investigation may be in progress already or can be set up. The data collected over time can then be used to estimate remaining life before repair becomes essential to maintain safety.[22-27]
- Where relevant as-built drawings do not already exist, preparation of as-built/ as-found drawings of sensitive or deteriorated structural areas based on the results of the Survey. Drawings should include updated information on modifications and repairs, e.g. rehabilitation of barriers or holes cut for services, and on deterioration found.
- Reporting to the Engineer and the Owner/Operator giving details of the observed condition of the structure, cladding and barriers, the potential implications for safety and future use, and proposing further actions including prompt pre-emptive maintenance where appropriate. The development of more serious and costly deterioration in the longer term arising from, for example, leaking drains and expansion joints, ponding seepage through cracks, breakdown of waterproofing, rust staining and spalling, can be reduced and/or delayed by giving priority to prompt pre-emptive maintenance. To avoid making deterioration worse and/or initiating it elsewhere, pre-emptive maintenance must be appropriate and correctly applied.

## 12.5 Special Inspection

12.5.1 Special Inspection by an Engineer may be necessary at any time to examine damage caused by an accident or reported structural distress, movement or inadequacy.

# 13. Structural Investigation

13.1     The Engineer may require the Condition Survey to be made more comprehensive by Structural Investigation, in particular through sampling of materials and tests to determine, for example, the reinforcement present and/or to diagnose the conditions within the structural components (Figure 13.1). The purpose is to determine details of the construction, and/or to confirm causes or extent of deterioration as part of the process of bringing together Appraisal of structure sensitivity and of damage due to deterioration.

Figure 13.1 Structural investigation in progress: taking concrete cores.

13.2 The Engineer should specify all investigative work, including locations where samples are to be taken. Sampling and testing may be undertaken by a materials test house or a materials and repair specialist.

13.3 Structural Investigation may be needed also to implement a Special Inspection.

13.4 The degree to which it is useful to extend the Condition Survey should be decided by the Engineer (or the Inspector in discussion with the Engineer as appropriate), bearing in mind the Owner's/Operator's requirements and the structural defects and deterioration found or suspected. Decisions will be needed on what tests to use, and when and where to use them, bearing in mind the precision required.

13.5 It is essential first to define clearly the purpose(s) of Investigation. In general, the purpose will be to determine the actual condition of the structure so that defects and causes of deterioration and distress can be diagnosed, a prognosis developed for the safety of the structure and, where needed, effective repair strategies identified. The aim of the Condition Survey of a concrete structure will usually be to identify those areas where there may be a structural defect, or deterioration is most advanced and critical or risk of reinforcement corrosion the greatest. It is essential to diagnose the causes and extent of deterioration before considering repair options.

13.6 The more common test techniques for testing concrete materials and components are briefly described in Appendix C. More comprehensive information on available techniques and on the planning and procedures of testing concrete structures may be found elsewhere.[30-32]

13.7 The relevant attributes should be measured. For example, an investigation that seeks only to find areas of minimum concrete cover to reinforcement in relation to corrosion risk may fail to identify bars with excessive cover that result in structural strength less than designed. The selection of measurements to be made should take into account any previous test data and the known or suspected extent of contamination and deterioration.

13.8 In general, direct observation or measurement of the factors in question is preferable to indirect methods. Direct observation or measurement is often not possible or practicable, however, and techniques that provide the information from which the parameter of interest can be estimated may need to be employed, e.g. measurement of ultrasonic pulse velocity (UPV) to provide an indication of concrete strength (see C2.3). In these circumstances, the reliability of the information obtained can be maximised by using two testing techniques in combination.[33]

13.9 When selecting test techniques for a particular Investigation, the Engineer should first define the information required, including criteria such as number of tests, level of accuracy, etc. The number of tests and level of accuracy should be no more than required for the subsequent Appraisal.

13.10 Investigation should be sufficient to establish reinforcement details where drawings are not available for elements that are potentially structurally sensitive. Deterioration effects may not be superficially visible for some structural forms and details. Special investigation procedures may be needed in these cases.

13.11 Condensation is usually most severe on the soffit of the top deck slab where, together with leakage through damaged or poor quality waterproof membranes, more corrosion of soffit reinforcement tends to occur compared with intermediate decks. De-icing salt is still commonly spread in car parks and it is also brought in on the wheels of vehicles. Chloride contamination is usually greatest in older car parks with external ramps as Operators have salted the ramps to keep the car park open in winter. Chloride contamination of the concrete can vary substantially from one area to another. For example, de-icing salt concentrations are likely to be greater on up-ramps than on down-ramps and in driveway areas. In addition, parking bays near to lifts and exits will be more frequently used than those in other areas. As a result, the wheel positions in these bays may be particularly at risk owing to ice and rain run-off from tyres. In general, therefore, the distribution of deterioration or defects in concrete car park structures is very uneven (Figure 13.2).

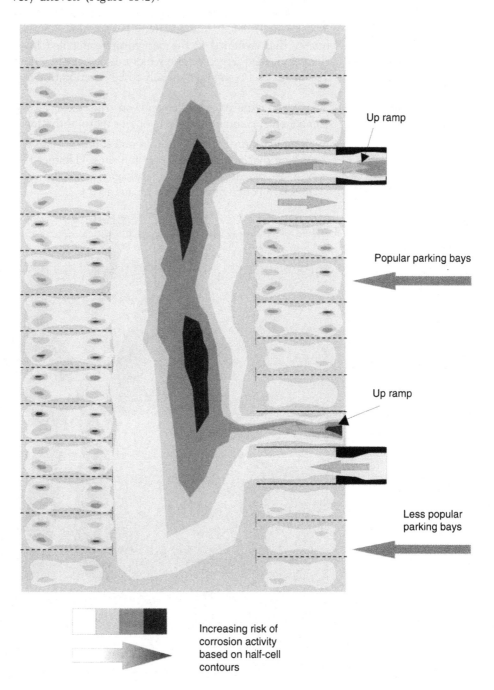

Up ramp

Popular parking bays

Up ramp

Less popular parking bays

Increasing risk of corrosion activity based on half-cell contours

*Figure 13.2 Schematic diagram of the results of a half-cell survey of a deck slab.*

13.12    For concrete structures, observed deterioration may indicate a local problem or it may be present widely in the structure. The amount of testing and the techniques used should be sufficient to enable sound interpretations of the structural condition, prognosis and repair needs.[22,31] A low level of testing can be a false economy. An unrepresentative view of condition may then be taken, resulting in repair works being overspecified or underestimated. In both cases higher repair costs are likely to arise ultimately.

13.13    A few random spot checks on cover, corrosion or core strength are unlikely to identify risks. Targeting investigation through sampling and testing in areas, i.e. ramps and running aisles, or in areas where defects and signs of early deterioration indicate poor quality will give a more accurate indication of risks. Where this investigation confirms deterioration, defect and/or strength variation indicating more variability or lower quality than assumed in the design, more detailed Investigation and Appraisal may then be needed. For concrete structures damaged by deterioration, testing of between 10 and 25% of components likely to be deteriorating may be needed.[20]

13.14    The tests required will depend on the local circumstances but, for concrete components, commonly include one or more of the following:

- Delamination survey of the tops and soffits of slabs to identify cracks and potential spalling that may become corrosion sites. Hammer tapping, chain dragging on tops of decks (Figure 13.3), or thermography can be used to locate delaminated and defective concrete cover and repairs (see C2.2).

*Figure 13.3   Chain drag impact survey.*

- Half-cell potential surveys to locate areas with the highest risk of corroding steel. The schematic diagram of a half-cell survey in Figure 13.2 clearly indicates the likely distribution of corrosion related deterioration through the variable distribution of chloride contamination by vehicles in a car park. The half-cell survey provides a clear warning of those areas most likely to be at risk of corrosion (see C3.3).
- Screening for chloride content in concrete components, especially on decks in areas of highest and lowest negative half-cell potentials, delaminated areas, traffic areas, parking bays and sheltered/exposed locations, to enable a complete 'picture' to be drawn of high/low risk areas of corrosion of reinforcement (see C3.2).
- Survey by local tests of concrete cover and carbonation depth. Tests at selected locations over the concrete surfaces are made to find areas where the depth of carbonation exceeds the depth of cover, placing the reinforcement at risk of corrosion (see C2.1 and C3.1). These tests can usefully be made at the locations sampled for chloride screening.
- Concrete coring for strength tests and/or pull-off tests to determine strength (see C2.4 and C2.5). Sufficient cores should be taken to establish variability of concrete strength and should include coring into areas showing indications of poor quality construction and/or deterioration.
- Petrographic examination of cores to assess the composition of the concrete, aggregate and cement types, water/cement ratio, cracking and possible sulphate attack and alkali silica reactivity (see C3.5).
- Exposure of reinforcement or prestressing tendons and measurement of amount of reinforcement present, the cover and the extent of corrosion. Cover measurements can also provide a calibration check of the cover meter survey (see C2.1). Prior to exposure of embedded steel, consideration should be given to any contractual or other warranties that may be invalidated by such action.
- For cladding, inspection of several fixings by remote viewing or mechanical exposure (see C4.1 and C4.2).
- For barriers, uncovering or dismantling a statistically representative sample of holding-down bolts for inspection of the bolts and the fixing detail and, in some cases, load testing (see C5.1). (Note: barriers fixed through asphalt are likely to be ineffective.)

13.15    Where extensive deterioration may be present in the structure, cladding or barriers, Structural Investigation may be needed annually over several years to enable the Engineer to confirm the causes of the degradation and its extent. This monitoring enables the rate of degradation to be tracked and provides a basis for formulating a deterioration model and updating predictions of residual service life. Procedures and techniques for monitoring structures are described elsewhere.[23–28] It is important in commissioning such monitoring work to recognize the need for coordination and continuity in gathering good quality information and hence in implementing the most cost effective Life-care Plan.

13.16    Samples and cores should be taken from non-critical structural locations of components as far as possible, and as agreed by the Engineer. Samples of materials (concrete, masonry, steel, etc.) for laboratory analyses can be obtained by a variety of drilling, sawing and cutting techniques. The inevitable damage caused by sample removal from the structure should be fully repaired in accordance with BS EN 1504.[19] Damage to reinforcement or prestressing tendons must be avoided.

13.17    Many non-destructive and partially destructive test techniques, e.g. pull-off tests, in fact, cause some damage or residual mark to the tested component, e.g. pull-off tests, even after making good and cleaning (see Appendix C). Owners and Operators should be informed of any likely effects on the appearance of the structure before the tests are made.

13.18  Testing should be undertaken by an organization working to a relevant quality assurance standard for the work required, e.g. to BS EN ISO9001 or UKAS accreditation providing assurance of sound practice.

13.19  Structural load tests may be required by the Engineer. Load tests may be useful in some circumstances to determine the capacity of fixings of cladding panels, of an edge barrier system or of holding-down bolts (see C5). Such tests can provide useful data where reliable calculation of capacity is not possible. For car park decks, load-carrying capacity can generally be calculated.

# 14. Structural Appraisal

## 14.1 Approach

14.1.1    A Structural Appraisal to evaluate the integrity and adequacy of the structure, cladding and edge barriers should be undertaken by the Engineer at intervals of not more than 16 years (see Table 5.1). Appraisal will generally be needed at intervals shorter than 16 years. For some older structures or structural parts, more frequent Structural Appraisal may be needed, in particular where the structure is contaminated by chlorides (even at a low level), highly carbonated or simply wearing out from vehicle movements.

14.1.2    The Structural Appraisal may be assisted by data from monitoring of the progress of deterioration at sensitive locations to confirm diagnoses of causes and to enable prognoses of future condition to be formulated.[22–27]

14.1.3    The Appraisal should be based on a review by the Engineer of all the information to hand from the Initial Appraisal and Condition Survey, including any Structural Investigation. The information should be sufficient to enable decisions on the extent and depth of Structural Appraisal needed. The purpose of the Appraisal is to assess the present condition, structural safety and adequacy of the existing primary structure, cladding and edge barriers against current requirements and to forecast future trends and needs for inspection and repair.

14.1.4    A partial or full Structural Appraisal may also be needed at short notice following a Special Inspection that indicates structural integrity may be in question, e.g. following the appearance of movement, or an accident. This Appraisal may also lead to a rescheduling of future Appraisals in the Life-care Plan.

14.1.5    On completion of an Appraisal, the Engineer should review inspection timescales and recommend to the Owner/Operator an appropriate schedule for the next Appraisal as part of an update of the Life-care Plan.

## 14.2 Principles and process

14.2.1    Practical approaches to Structural Appraisal of structures are usually based on modifying design models, possibly associated with more refined structural analysis, and using measured inputs for structural and deterioration parameters; see for example approaches used for highway bridges.[34] Decision support tools and reliability-based approaches have been recently developed[24] and are providing a useful addition to asset management.

14.2.2 The Structural Appraisal should enable the Engineer to assess structural inadequacies and progressive deterioration in the structure, its claddings and barriers. The principles given elsewhere should be followed.[28,35] The process is generally cyclical in nature (see Figure 12.1). Comparison of as-built strength with current requirements and reductions in strength due to deterioration are generally the main focus.

14.2.3 The principal steps in the process of Structural Appraisal of the structure depend on the requirements of the Owner/Operator. For concrete structures, they are generally along the following lines:

(1) Establish the original basis of design and levels of safety.
(2) Calculate the capacity as-built, using as far as possible measured values for stiffness and strength, e.g. tests on concrete cores.
(3) Compare (1) and (2) with current requirements.
(4) Consider actual live loading, compared with design loading.
(5) Consider possible secondary effects, not treated explicitly in the original design, e.g. possible shrinkage or temperature effects, significant construction tolerances.
(6) Re-analyse the structure, using modified values for member sizes, stiffness and strength as measured and modified for the effects of deterioration, e.g. section loss of reinforcement bars. This re-analysis is to assess the possible redistribution of forces and moments, while taking account of (4) and (5) above. It is also to identify possible weaker load-bearing mechanisms induced uniquely by deterioration.
(7) Calculate the effects of deterioration at each critical section as determined from (6) above, for each relevant action effect (bending, shear, etc.), singly or in combination as appropriate, using the corresponding design equations modified for the purpose.
(8) Establish criteria for minimum acceptable technical performance.[25,28]
(9) Compare outputs from (3), (6) and (7) with the criteria.

14.2.4 The comparisons provide the bases for evaluation of options, including those on the nature and urgency of repair or strengthening interventions. Similar principles apply to the Structural Appraisal of cladding or edge barriers.

## 14.3 Aspects of appraisal

14.3.1 For the structure, Structural Appraisal should usually include:

(1) Appraisal of the serviceability and strength of the structure in its current condition and in a future more deteriorated condition over its residual life,[25] taking into account test results and data from any monitoring.
(2) Appraisal of potential to fail suddenly at any local structural detail, e.g. slab shear strength at columns, corbels supporting beams, or area of stress concentration especially in precast concrete deck systems.
(3) The vulnerability of details, e.g. corbels, to deterioration over time and the potential for accelerated deterioration causing loss of capacity leading to sudden failure.
(4) The robustness of the primary structure to instability and resistance to progressive collapse at present and as identified local deterioration progresses, including the situation where a vehicle strikes a vulnerable key element.
(5) Appraisal of movement and construction joints, especially where they have failed or leakage of contaminated water has occurred through them.
(6) Appraisal of damaged areas and areas where repairs have been carried out previously.
(7) For basement car parks, appraisal of any implications for the building above.

14.3.2 For the cladding, Structural Appraisal should usually include:

(1) Appraisal of the adequacy of fixings.
(2) Appraisal of the integrity of the cladding and its residual life.
(3) Whether, over the residual life, further deterioration might lead to unsound cladding elements liable to present a hazard to safety.

14.3.3 For edge barriers, Structural Appraisal may usually be approached in stages as follows:

(1) Determine the types of edge barrier existing, their condition, dimensions and fixing detail on the basis of the Condition Survey and any subsequent Structural Investigation, e.g. checks of whether holding-down bolts have worked loose and exposure to check condition.
(2) Evaluate adequacy for functions of vehicle restraint and pedestrian safety by reference to current and proposed standards (see A4 and C5), and likely deterioration over the next 8 years. Consider the need for an installation compliance test and implement if appropriate.[4]
(3) Identify priorities for improvement/replacement.
(4) Evaluate costs of preferred options for improvement/replacement of barriers where needed.

## 14.4 Discussion

14.4.1 Overall structural stability (including vulnerability to progressive collapse), flexural capacity of slabs, punching shear at column/slab connections and vehicle restraint should be examined. Published guidance on structural appraisal,[28] together with that given in current design codes, provides more detailed advice.

14.4.2 There may be deficiencies in current design codes when used for Appraisal. Current design codes, whilst providing a conceptual basis for appraising as-built strength, may not cover the structural configurations and details found in an existing car park structure. Design codes cover particular ranges of structure and structural form. The ranges are not generally fully defined. Where a structural design is innovative relative to the assumptions in the code, e.g. variable spans compared with a code based on uniform spans, detailed examination using engineering first principles may be required. Literature research or physical testing of the structure may be needed. The performance of the structure during its life is likely to provide evidence of the examination required.

14.4.3 It is important to recognize therefore that the code(s) to which an existing structure, cladding or edge protection was designed may, in some circumstances, not provide a reliable basis for determining current strength and robustness. Checks should be made of changes in design standards and the 'state of the art' relating to the particular structural form. They may indicate structural features that have low reserves of strength compared with modern standards of adequacy, e.g. relating to horizontal forces, braking forces, barriers, progressive collapse, shear strength, ineffective propping and repairs.

14.4.4 Codes of practice for structural design include factors of safety assuming that construction will be within specified tolerances and workmanship. When a Condition Survey indicates that the construction is outside assumed tolerances, detailed structural checks should be made. Examples of out-of-tolerance construction found in some car park structures include cover to reinforcement, position of supports, reinforcement of bearings of beams on corbels and ledges, weak concrete as-built or deteriorated, and spans between supports for edge barriers greater than specified.

14.4.5 Checks of the original design calculations relating to sensitive elements and details can provide useful background, especially where a now-obsolete code was used. The checks may find that the original design calculations contained errors or that difficult-to-construct design details were used. Such findings provide pointers for further investigation. It may also be valuable to make assessments of capacity, serviceability and durability based on tests and research data where the construction is not in accordance with a recognized design standard.

14.4.6 Construction defects and concrete deterioration identified in the Condition Survey need to be considered explicitly in the Structural Appraisal. Defects can be present owing to incorrect spacing, curtailment and anchorage of reinforcement. The reinforcement drawings may not have reflected the design intent and/or the steel fixing may not have replicated the details on the drawings.

14.4.7 The Appraisle may take into account the increase in the compressive and tensile strength of the concrete that will happen over the years, and also the weakening of components through corrosion and other forms of deterioration and through patch repairs. It should also be appreciated that the steel strength used in the design was probably based upon the 95% characteristic strength appropriate to the grade of steel generally available on the market at the time. It may be possible to justify an increased strength for use in Appraisal by testing samples of the steel taken from the structure (see C2.6).

14.4.8 Assessments of current remaining capacity and its likely further reduction over time are then needed. Assessments of capacity and ductility of sensitive details whose failure might be sudden with the potential to trigger progressive collapse should also be made (see Appendix D). Particular caution is required in considering bond and anchorage of reinforcement in deteriorated reinforced concrete. They may be more critical than for the original design.

14.4.9 The work should identify any needs for immediate repair, and lead to the determination of options for the future maintenance of the structure, taking into account foreseen patterns of deterioration.[25] More than one destructive mechanism may be acting simultaneously, e.g. freeze/thaw and reinforcement corrosion, and it is necessary to take account of combined effects.[25] A serious situation in terms of structural safety and the residual life of a concrete structure may arise if corrosion of reinforcement, perhaps the most common form of deterioration, cannot be stopped or inhibited economically at structurally critical locations. The proposed method of repair should aim to prevent the electrochemical process of corrosion from continuing or occurring elsewhere.

14.4.10 Calculations of remaining capacity of deteriorated components should be framed with caution. Usually partial factors can be reduced, for the same overall reliability, since measurement will have provided better estimates of dimensions, loads and material properties than those assumed in design.[28] Such back-analysis for the undeteriorated condition is usually reliable, but caution is needed in including allowance for significant deterioration that is present since quantifying the extent and significance of deterioration cannot be precise.

14.4.11 The presumption that partial factors for design can be reduced in Structural Appraisal can be incorrect in some situations. Increased factors may be needed where variability or effects covered by partial factors are greater than assumed in the design code. In particular, increases may be needed where there is sensitivity to construction tolerances, detailing requirements are not met, effects of deterioration are uncertain and/or the consequences of local failure would be serious, for example at some corbels. Alternatively it may be better to evaluate the implications of the deterioration explicitly (see Appendix D). Note that design codes for concrete have no margin or factor to cover loss of strength from deterioration.

14.4.12  Where secondary influences not covered in the original design are significant, e.g. erection tolerances or shrinkage effects, they should be treated separately in an Appraisal [see 14.2.3 (5) above]. In these cases, simply keeping the same, or increased, partial factors for traditionally considered loads may not be adequate to take these influences into account.

14.4.13  Estimates of future rates of deterioration may be possible, especially where monitoring over time has provided indications of current rates.[22-27,36] Where there is a basis for considering estimates of future rates to be reasonable, back-calculation can be made to predict when the minimum acceptable technical performance will be reached. Such calculations provide useful input into planning future inspections and repair interventions. However, caution is needed as back-analysis of deteriorated structure relies largely on engineering judgement.

14.4.14  An essential part of the Structural Appraisal of a car park structure is the evaluation of susceptibility to progressive collapse. Most forms of car park structure as built are not vulnerable to progressive collapse. Some structures, however, may be vulnerable if local damage, e.g. an accidental or deliberate impact by a vehicle on a column, or local structural failure, e.g. punching shear at a slab support, should occur. Vulnerability may increase while the structure is in use as a result of deterioration of materials and components. A structure may also become more vulnerable as a result of insufficient propping and poorly executed repair works, and loss of cross-sectional area of reinforcing bars. Evaluation of susceptibility to progressive collapse is discussed further in Appendix D.

14.4.15  The results of the Structural Appraisal should be summarized. Assessed current capacities, compared with those required at all critical locations, should be included. Assessment of future rates of deterioration and the implications for the safety and use of the car park should be given and the effectiveness of waterproofing and other measures to slow deterioration should be examined. These outcomes then provide the Engineer with a basis for recommending options for repair and rehabilitation or replacement. Interim measures to maintain safety pending remedial works may be required in some cases.

## 14.5 Summary of principal actions in a typical Structural Appraisal

14.5.1  Actions should be taken as scheduled by the Engineer. They will normally include:

- Completion of appraisal calculations for the structure and production of a detailed summary of the results with assessed structural capacities (as-designed, as-built and as-found deteriorated) being compared with those required by design standards for all critical locations. The accuracy of as-found deteriorated capacities may be low given the subjective judgements required in the calculations. These calculated values should therefore be used with caution. In some circumstances appraisal may be based on judgement only without calculation, provided the basis of the judgement is given.

- The summary should identify any particular weaknesses, lack of robustness or potential for progressive collapse. Standards applicable at the time of construction are generally appropriate except where the structural configuration is not within their scope or clauses have been withdrawn subsequently for technical reasons, e.g. empirical rules for flat slab design. For shear in flat slabs, and Lift-Slab structures in particular, current standards should be used.

- Appraisal of the safety and integrity of cladding.

- Appraisal of the adequacy of edge protection and fixings for vehicle restraint by reference to published performance data,[3,4] or testing where their adequacy cannot be determined with confidence. Appraisal of adequacy for protection of pedestrians should also be considered.

- Appraisal of the implications of trends in deterioration on the assessments above and the risks of accelerated deterioration, especially for details that are not easily inspected or are critical to structural safety. Where deterioration mechanisms are weakening the structure, criteria for minimum acceptable technical performance should be developed and applied bearing in mind the limitations in reliability of the condition assessment. The next Condition Survey and repair and strengthening works should be undertaken before minimum acceptable technical performance is reached.

- Arranging for monitoring of sensitive deteriorating parts of the structure, e.g. corbels or integrity of steelwork connections.

- If the structure is deficient in strength or unacceptably sensitive to partial or progressive collapse, provision of costed options for interim measures, pending remedial works, together with costed options for remedial works.

- Where the structure, cladding and/or edge barriers are unsafe, inadequate or substantially deteriorating, provision of costed options for interim measures and remedial works.

- Production of costed options for the long-term inspection, preventive and capital maintenance and periodic re-appraisal of the structure, cladding and barriers with specific reference to ensuring continued structural safety and dealing with deterioration that may not have safety implications.

- Reporting to the Owner/Operator, giving the results of the Appraisal and proposing the future Life-care Plan, including options for future inspection, maintenance, repair and rehabilitation or replacement.

# 15. Maintenance, repair, rehabilitation and replacement

## 15.1 Objective

15.1.1 The primary objective of maintenance, repair, rehabilitation or replacement is generally to maintain the structure, cladding and edge barriers in a safe and serviceable condition at reasonable residual-life cost compatible with the Owner's/Operator's requirements as reflected in the Life-care Plan. All proposals for repair should be subject to the advice of the Engineer.

## 15.2 Routine maintenance

15.2.1 The primary purpose of routine maintenance is to keep a multi-storey car park clean and fit for use. The most important agents promoting deterioration of car park structures are de-icing salts and water brought into the structure by vehicles and the weather. A secondary purpose of routine maintenance is therefore to minimize the deteriorating effects by keeping the environmental conditions within the car park as dry as possible and reducing the amount of contamination, i.e. maintaining good drainage, clean decks and falls. Poor drainage, blocked drains and ponding on slabs due to construction defects or creep deflection enhance freeze/thaw and traffic damage to deck surfaces. Unsealed deck surfaces allow seepage of water and chlorides into the underlying structure through joints, cracks and porosity in poorly compacted concrete, especially where ponding occurs.

15.2.2 A ban on the use of de-icing salt in the car park will reduce the risk of chloride-induced corrosion although salt will still be imported onto the decks by vehicles in the winter months. Improved maintenance procedures may be necessary where, for example, drains are repeatedly blocked or ponding occurs. Such measures are likely to enhance the life of structural components before repair or rehabilitation is needed.

15.2.3 Routine maintenance should include:

- Keeping the structure as dry as practicable by keeping drains unblocked and clear of debris. Experience should dictate how frequently drains are cleaned. Known points prone to blockage should be identified for frequent checking and clearance, and for possible preventive action.

- Removing loose concrete debris from decks arising from traffic damage, frost or other influences.
- Regularly washing down decks and ramps, especially during the winter months and at the end of the winter, to remove as much de-icing salt and standing water as possible and to flush through drains. Washing down is unlikely to reduce the rate of deterioration significantly unless decks are waterproofed first.
- A ban on the use of chloride-based de-icing materials in the car park.
- Minor repairs to stop leaks from drainage systems, and to reinstate barriers damaged by accidental vehicle impact.
- Where possible, cleaning surfaces defaced by vandalism and making good minor damage to surfaces.

15.2.4    Hazards from falling spalled concrete should be avoided by giving maintenance priority to the regular removal of loose surface concrete, especially in and over public areas.

15.2.5    The Operator's employees usually undertake routine maintenance. Responsibility for minor repairs, e.g. to drains, barriers, joints or waterproofing, may be placed with a third party who should be properly supervised. Those undertaking routine maintenance should be trained appropriately to minimize the risks of incorrect work that may make matters worse.

## 15.3  Protection, repair and rehabilitation or replacement

15.3.1    The protection and repair of concrete structures require complex design and specification work. This work is not always adequate and, as a result, concrete repairs are often not as durable as they could be particularly where chlorides are present.[37] The basic principles for protection and repair of concrete structures subject to corrosion damage, or which are expected to need such measures to minimize future damage or deterioration, are given in BS EN 1504 and elsewhere.[19–21,38] The objectives and basis for the choice of protective and repair products and systems for concrete structures in general are given in the BS EN 1504.[19] Updates on current practice and standards may be found in recent reports from the Department of Trade and Industry project 'Corrosion in Concrete'.[36]

15.3.2    Car park structures are one of the few types of structure where live loads are in direct contact with the structure, which 'wears out' in heavily trafficked areas. They are also subjected to vibration and impact, especially if ramps are steep. These factors make sound repair extremely difficult.

15.3.3    Patch repairs may be appropriate in some circumstances (Figures 15.1 and 15.2). Significant repairs to a concrete structure involving breaking out and patching sizeable areas of concrete or surface treatments for exposed reinforcing bars and replacing concrete cover should not be considered routine maintenance. Such repairs may have structural implications and should be treated as structural repairs. They should be clearly specified in terms of purpose, type and methodology.

15.3.4    Protection and repairs to deteriorated concrete structures all too often are less durable than expected.[21] Failure of a protective treatment or repair may occur prematurely for many reasons, in particular inadequate or inappropriate specification, materials

*Figure 15.1 Application of patch repair material.*

*Figure 15.2 Application of patch repair material.*

and workmanship. Full diagnosis of the causes and extent of deterioration and the structural consequences is needed before preparing a repair specification and contract to satisfy the Owner's/Operator's requirements in terms of both life expectancy and costs.[19-21]

15.3.5   In developing proposals for the protection and repair of concrete car park structures (and for updating the Life-care Plan), the Engineer should consider the options in the light of the results of the Structural Appraisal, requirements for traffic management and the costs of implementation. The resulting proposal should meet the Owner's and

Operator's requirements and may involve one or more of the following considerations:

- Structural inadequacies identified in the primary structure, its cladding or barriers may imply a level of safety below that expected by current standards. Where this is the case, the Engineer may recommend that works to strengthen or replace the defective structural features should be undertaken as soon as possible, and where necessary protective measures to safeguard the public be implemented in the meantime.
- When defining rehabilitation works, it is essential for the Engineer to understand the design of the structure, the need for propping/through propping, and the critical role of the bond between the repair and original material. For decks with thin concrete sections, total deck replacement should be considered as an alternative to repair (Figure 15.3).
- Where deterioration of the structure is advanced, replacement may be a more cost-effective option than retaining the existing structure with continuing high maintenance costs. In the meantime measures may be needed to control the degradation and possible hazards, e.g. falls of spalled concrete.
- Where deterioration, corrosion, spalling or cutting out exposes reinforcement, it must be considered as potentially reducing structural adequacy and as such requiring advice from the Engineer.
- The strength of the component after repair in some critical areas can be very sensitive to adhesion at the interface between original concrete and the repair material. After repair the component may be prone to a sudden mode of failure as the interface adhesion breaks down under load. Inappropriate concrete repair materials, stronger than the original material, tend to be specified.
- The achievement of a sound concrete repair is critically dependent on reliable workmanship on site. In particular, cutting out should avoid damage to the substrate as far as possible and the repair material should be well compacted.
- Local repair and coating can in some circumstances be based satisfactorily on the results of visual inspection.[19–21] Repainting of structural steelwork can often be successfully scheduled by visually checking the condition of protective coatings. For reinforced and prestressed concrete elements, an approach consisting of repeated

*Figure 15.3 Removal of deteriorated deck slab prior to replacement.*

patch repairing of cracked concrete cover followed by cosmetic improvement may not be cost-effective or, more importantly, may not maintain safety. If no waterproofing is applied as part of the remedial works, patch repair is unlikely to slow down overall deterioration due to reinforcement corrosion where significant levels of chloride are present in the concrete. Rather, corrosion can accelerate local to the patch.[21,22,30] In general cosmetic repairs and coatings may have a short life and not be cost-effective where corrosion or other underlying deterioration processes are present and not identified and inhibited. Only where chloride contamination is not significant and carbonation of low cover to reinforcement has caused local spalling, e.g. of concrete cladding, is treatment by a protective coating after removal of loose surface material likely to be cost-effective.

- Preventive maintenance and repair can be undertaken to control the rate of deterioration and so prolong service life. For concrete structures, protective coating treatments may be used to reduce the ingress of moisture and chlorides. Such treatments can be cost-effective particularly if applied in the early years of life of a car park. Where reinforcement corrosion is already occurring, it can be treated and brought under control by the electrochemical technique of cathodic protection as recommended in BS EN 1504.[19] The protection can be targeted for application to critical areas and can be monitored and adjusted throughout the life of the structure. In newly built structures it can be economical to provide electrical continuity of the reinforcement and apply cathodic prevention prior to deterioration occurring or achieving damaging effects. Other electrochemical processes, including chloride removal and re-alkalization, can be used to treat contaminated concrete and extend service life. More recent additions to the suite of available techniques are surface applied corrosion inhibitors and electrokinesis. Specialist advice on the applicability of these techniques should be sought and is available from the industry trade association (see Section 20). In some car park structures, cracking due to alkali silica reaction (ASR) in concrete has been found. Guidance on determining residual life and the management of ASR affected structures is available.[25,39]

- To contribute effectively to the structural strength of the component, a concrete repair must include reinforcement and/or mechanical interlock at the interface between original concrete and the repair material. Punching shear anchorage zones are particularly difficult to repair reliably. Full depth recasting or another technique is likely to be required once effective anchorage of the reinforcement has been damaged (see D4.3).

- Load sharing between original concrete and repair material and the ability to sustain stress concentration on completion (Figures 15.4 and 15.5) are sensitive to structural support during repair hardening.[40]

- The Engineer should design all propping taking into account the structural limitations of the existing structure. Where top steel has failed, the main span may need to be jacked up and propped to relieve the stress and to enable continuity of the top steel to be achieved within the repair.

- Concrete repairs may fail by delamination when initial thermal and shrinkage strains during casting and curing and physical mismatch between original concrete and repair material alter stress distributions. This possibility needs explicit consideration when the repaired location does not have a ductile failure mode when under load. Guidance may be found in BS EN 1504.[19]

- Appropriate principles should be adopted relative to the defects found in the concrete.[19-21] Repairs to concrete decks may require propping prior to cutting out to prepare for the application of new material. Before repairs to decks or cladding panels, safety should be assessed for all stages, i.e. as built, as deteriorated, as cut out, as repaired and with repair delaminated. Care is needed especially with prestressed concrete construction.

Figure 15.4 Column/slab connection in a Lift-Slab structure during preparation for recasting.

Figure 15.5 Replacement of extensive areas of deck slab.

- Rehabilitation should provide as far as practicable efficient drainage systems for decks and no interruptions to drainage paths by joints or kerbs. Where ponding occurs on decks, additional drainage should be installed when practicable. Flow of surface water over joints should be avoided if possible. Decks may be waterproofed. Waterproofing of concrete decks that have no falls is unlikely to be reliable. For concrete top slab decks, where rain, freeze/thaw and temperature changes are at their most severe, waterproofing is needed to extend concrete durability. The provision of a protective lightweight roof can be advantageous.

*Figure 15.6 Edge protection barrier after vehicle impact.*

*Figure 15.7 Edge protection comprising flexibly mounted, hot-rolled steel posts with motorway barrier section rails after restraining a simulated vehicle impact during successful system compliance tests.*

- Analyses of rates of deterioration in concrete structures can provide a basis for delaying repair until the optimum time taking into account commercial factors.[15,25]
- For edge protection barriers, upgrading may be necessary, or replacement may need to be planned where damaged (Figure 15.6) or if deterioration is advanced. Design of replacements should take account of load transfer to the original structure, barrier deflection under vehicle impact and parking bay length (Figure 15.7).
- The structure repair specification, with the extent of cut out and repair and support requirements marked up on reinforcement drawings of each repaired area, should be prepared including a method statement (Figure 15.4).
- The repairs should be checked during installation and certified by the Engineer on completion.
- Appropriate records of repairs, e.g. specifications and locations, should be kept to assist in future Structural Appraisal.

15.3.6  For steel-framed structures, proposals for protection and repair may need to address: corrosion of steelwork and metal decking, prising of joints, impact damage, excessive deflection, reflective cracking of prestressed concrete units above beams and monitoring of bearing details.

# 16. Health and safety of personnel on site

## 16.1 General

16.1.1 Requirements for health and safety should be met for all inspection, investigative, maintenance and repair work on site. Appropriate assessments of risks should be carried out (see 5.3.11). Inspections should not be carried out by inexperienced people working on their own. An Inspector/Engineer, however experienced, visiting a site alone, should report back at specified intervals that all is well.

16.1.2 Those involved in work on site should be made aware of likely hazards and the work planned to minimize personal risks. Safe systems of work and appropriate safety equipment should be used. Particular care should be taken when working at height or in confined spaces.

16.1.3 Multi-storey car parks are public spaces and the safety of the public should be the prime concern. All work on site should be arranged so that it does not constitute a hazard to the public.

## 16.2 Common hazards

16.2.1 The following list of potential site hazards associated with site inspection or investigations is not exhaustive but gives preliminary guidance.
- Most inspections of car park structures take place whilst the car park is operational. Lighting may be poor. Hazards can arise from moving vehicles driven by car park users.
- The majority of accidents on construction sites generally result from falls from access equipment. Falls of people, e.g. from untied ladders, and objects falling from scaffolds cause most accidents. Inspection of car park structures that may be in use can present similar risks. Specific training is required for the safe erection and use of scaffolds, mobile towers, powered access equipment and rope access techniques.
- Biological hazards can arise from accumulations of rubbish, organic waste and vermin, particularly in confined spaces such as stairwells and lift shafts. Used hypodermic needles are a hazard sometimes found in stairwells, taped to handrails or jammed into lift control buttons. Such hazards should be removed by appropriate specialists on a regular basis as a part of routine maintenance.

- Permanent formwork made of asbestos cement is a common hazard. If in any doubt about the presence of asbestos, specialist advice should be sought prior to disturbing any permanent formwork or other possible asbestos material. Contact with the local authority Environmental Health Department is essential.
- Discarded petrol tanks are often found in basement and ground floor areas.
- Confined spaces such as manholes and service ducts can present a danger of oxygen-deficient atmospheres. They should not be entered by anyone alone. Care should be taken to avoid the health hazards of contaminated water, e.g. Weil's disease.
- Fire or impact damage to a car park structure may make it unsafe to carry out a full inspection before temporary support and protection are provided.
- Live electricity cables can be a hazard if struck during investigation work. Buried redundant cables should never be assumed to be dead. The supply of electricity to equipment used for inspection and investigation should be properly protected to enable safe use.
- Where spalling concrete has been identified, protection for the public and those carrying out site work may be needed by use of, for example, debris netting or restrictions on access.

16.2.2　Condition Surveys and Structural Investigations may generate hazards requiring protective and control measures, for example:

- debris netting and/or banksmen below areas where loose/spalling concrete is being removed by hammer impact survey;
- a 'crash' deck to protect areas below sites of load testing of precast claddings.

## 16.3 Personal protective equipment

16.3.1　Daily Surveillance and Routine Inspections of multi-storey car parks by on-site staff should be carried out from points of public access. Personal protective equipment is not normally required but care is needed to avoid danger from moving vehicles.

16.3.2　Those undertaking Condition Surveys and Structural Investigations should be provided with personal protective equipment suitable for the task in hand. Safety helmets and protective footwear may be needed, particularly when low headroom presents risks of injury to heads by collision with slab soffits. Projecting reinforcement, tying wire, nails, etc., protruding from the surface of the concrete can be a hazard. Goggles for eye protection should be worn for drilling, sawing, breaking out or similar work. Ear protection and breathing masks may be needed when some equipment is in use. Appropriate protection should be provided to those working at height.

16.3.3　During surveys, investigations, repair and refurbishment works, protective measures for site personnel may be needed associated with, for example:

- the continuing access to parts of the car park by vehicles and the public. Traffic management provision may be required;
- electrical and other services that are buried within the structure;
- dust and vibration from tools;
- the use of solvents, paints, etc;
- maintaining safety of the structure, e.g. by propping before making structural repairs.

# 17. Reporting and records

## 17.1 Inspections

17.1.1 The Inspector and Engineer should report findings in response to their briefs received from the Owner or Operator.

## 17.2 Condition Survey

17.2.1 In general the Inspector's report on a Condition Survey should cover the condition of the primary structure, its claddings and barriers, and the structural defects and forms and extent of deterioration found. The Condition Survey should refer to the Life-care Plan and include review of the inspection and maintenance schedule and the long-term frequency of, and method of inspection for, each type of element. The report should include photographs and condition should be recorded on survey drawings.

## 17.3 Structural Investigation

17.3.1 The report of a Structural Investigation should give a description of the investigation required, the methods used, the reasons for their choice, details of the investigation carried out and any limitations. The results should be presented in a form, as discussed with the Engineer, that enables them to be taken into account in a Structural Appraisal.

## 17.4 Structural Appraisal

17.4.1 The Engineer's report on a Structural Appraisal should summarize the findings on structural adequacy, integrity, deterioration and safety, and on the residual life of the structure, cladding and edge barriers. The report should be expressed in terms that can be understood by a layperson.[28]

17.4.2 Recommendations for any immediate action needed to provide public and structural safety should be given.

17.4.3    The report should describe the options for future management of the car park structure relative to its required life and the Owner's/Operator's requirements. Management possibilities including combinations of repair and rehabilitation options and the timing of their implementation should be discussed. The report should highlight what must be done, what could be done and what can wait. The recommended options should be costed. Revisions to the Life-care Plan should be proposed.

## 17.5  Records

17.5.1    In addition to forwarding reports to the Owner/Operator as requested, copies should be passed for safe keeping into the record of the car park's history (see 5.4).

# 18. Main Recommendations of Part 2

18.1　The Engineer, the professional engineering adviser to the Owner/Operator, is recommended for each car park structure to:

(1) review current arrangements for inspection and maintenance in the light of these Recommendations;

(2) advise on getting started on a continuous life-care process in accordance with these Recommendations;

(3) where current inspection and maintenance arrangements are not to the standards of these Recommendations, recommend to the Owner/Operator the changes needed.

# 19. References to Parts 1 and 2

1. *Structural Safety 1994–96: Review and recommendations.* Standing Committee on Structural Safety, 11th Report, January 1997.
2. *Design recommendations for multi-storey and underground car parks.* 3rd edn. Institution of Structural Engineers, London, 2002.
3. *Edge protection in multi-storey car parks – design, specification and compliance testing.* DETR Partners in Innovation Contract 39/3/570 cc 1806. Report, October 2001.
4. *Edge protection in multi-storey car parks – assessment method for installed restraint systems.* DETR Partners in Innovation Contract 39/3/570 cc 1806. Report, October 2001.
5. *Enhancement of whole life performance of existing and future car parks.* Report of DTLR Partners in Innovation Project, 2002.
6. *Structural Safety 1996–99: Review and recommendations.* Standing Committee on Structural Safety, 12th Report, February 1999.
7. *Pipers Row car park, Wolverhampton: quantitative study of the causes of the partial collapse on 20 March 1997.* Report, Health and Safety Executive (in preparation).
8. Turnbull, N. et al. *Internal Control: Guidance for Directors on the Combined Code.* Institute of Chartered Accountants in England and Wales. September 1999.
9. *The Building Regulations 2000,* SI 2531.
10. *Building Act 1984.* HMSO, London.
11. Health and Safety at Work etc. *Act 1974.* HMSO, London.
12. *Workplace (Health, Safety and Welfare) Regulations 1992.*
13. *Occupiers' Liability Acts, 1957 and 1984.* HMSO, London.
14. *Buildings and constructed assets – service life planning.* ISO 15686. British Standards Institution, London.
15. *Whole life assessment of highway bridges and structures.* Highways Agency Advice Note BA81 (Draft), 2002.
16. *Management of Health and Safety at Work Regulations 1999 and Approved Code of Practice.* HMSO, London.
17. *Construction (Design and Management) Regulations 1994,* SI 1994 no. 3140, HMSO, London, 1995, and *Approved Code of Practice.*
18. *Control of Substances Hazardous to Health Regulations.* HMSO, London, 1994.
19. *BS EN 1504 Series: Products and systems for the protection and repair of concrete structures.* British Standards Institution, London.
20. Currie, R. J. and Robery, P. C. *Repair and maintenance of reinforced concrete.* Building Research Establishment Report BR 254, 1994.
21. *Corrosion in steel and concrete. Part 1: Durability of reinforced concrete structures. Part 2: Investigation and assessment. Part 3: Protection and remediation.* BRE Digest 444, February 2000.
22. *Testing and monitoring the durability of concrete structures.* Concrete Bridge Development Group Guide No. 2. The Concrete Society, 2000.
23. Matthews, S. L. Deployment of instrumentation for in-service monitoring of structures. *Structural Engineer,* 2000, 78, No. 13, pp. 28–32.

24. Moss, R. M. and Matthews, S. L. In-service structural monitoring – a state of the art review. *Structural Engineer*, 73, No. 2, pp. 23–31, 1995.

25. CONTECVET: *a validated user's manual for assessing the residual service life of concrete structures: Manual for assessing concrete structures affected by ASR; Manual for assessing concrete structures affected by frost; Manual for assessing corrosion-affected concrete structures.* EC Innovation Programme, IN309021. British Cement Association, 2002.

26. Corrosion of reinforced concrete: electrochemical monitoring. *BRE Digest* 434, November 1998.

27. Management, maintenance and strengthening of concrete structures. *Fib Bulletin* 17, April 2002.

28. *Appraisal of existing structures*. Institution of Structural Engineers, October 1996.

29. *Durable post-tensioned concrete bridges*. Technical Report 47, 2nd edn. The Concrete Society, 2002.

30. Bungey, J. H. and Millard, S. G. *Testing of concrete structures*. 3rd edn. Blackie Academic and Professional, 1996.

31. *Diagnosis of deterioration in concrete structures – identification of defects, evaluation and development of remedial action*. Concrete Society Technical Report No. 54, 2000.

32. Kay, T. *Assessment and renovation of concrete structures*. Longman, 1992.

33. Robery, P. C. and Casson, R. B. J. Investigation methods using combined NDT techniques. *Construction Repair*, November/December 1995, pp. 11–16.

34. *Design Manual for Roads and Bridges: The Assessment of Highway Bridges and Structures – BD21/97; The Assessment of Concrete Highway Bridges and Structures – BD44/95 and BA44/9; The Assessment of Steel Highway Bridges and Structures – BD56/96 and BA56/96.* The Stationery Office, London.

35. *Strategies for testing and assessment of concrete structures*. CEB Bulletin 243. Fédération Internationale du Béton, 1998.

36. Corrosion in Concrete. Reports. Trend 2000 Limited. Report 1 – *Review of NDT Survey Techniques*; Report 2 – *Handbook for Corrosion Rate Measurements*; Report 3 – *Corrosion Rate Measurement*; Report 4 – *Guide to the Maintenance, Repair and Monitoring of Reinforced Concrete Structures*; Report 5 – *Evaluation of Life Prediction and Modelling*; Report 6 – *Protection Strategies for the Design of New Structures against Reinforcement Corrosion*; Report 7 – *European Standards Affecting the Maintenance & Repair of Reinforced Concrete Structures*; Report 8 – *Standards for Specification of Concrete with Reference to Avoidance of Reinforcement Corrosion*;

37. *Field Studies of the Effectiveness of Concrete Repairs. Report.* Health and Safety Executive (in preparation).

38. Monographs: 1. *Reinforced concrete – history, properties and durability*; 2. *An introduction to electrochemical rehabilitation techniques*; 3. *Cathodic protection of steel in concrete – an international perspective*; 4. *Monitoring and maintenance of cathodic protection systems*; 5. *Corrosion mechanisms – an introduction to aqueous corrosion*; 6. *The principles and practice of galvanic cathodic protection for reinforced concrete structures*; 7. *Cathodic protection of early steel framed buildings*; 8. *Cathodic protection of steel in concrete – frequently asked questions*. Corrosion Protection Association.

39. *Structural effects of alkali-silica reaction*. Institution of Structural Engineers, London, July 1992.

40. Canisius, T. D. G. and Waleed, N. *The behaviour of concrete repair patches under propped and unpropped conditions – critical review of current knowledge and practice*. FBE Report 3, CRC, London, March 2002.

# 20. Useful web sources

| | |
|---|---|
| Association for Structural Engineers of London Boroughs | www.aselb.com |
| Association of Civil Engineers | www.acenet.org.uk |
| British Consultants Bureau | www.bcbforum.demon.co.uk |
| British Parking Association | www.britishparking.co.uk |
| Concrete Society | www.concrete.org.uk |
| Construction Industry Research and Information Association | www.ciria.org.uk |
| Corrosion Prevention Association | www.corrosionprevention.org.uk |
| Health and Safety Executive | www.hse.gov.uk |
| Institute of Chartered Accountants in England and Wales | www.icaew.co.uk |
| Institution of Civil Engineers | www.ice.org.uk |
| Institution of Engineers of Ireland | www.iei.ie |
| Institution of Structural Engineers | www.istructe.org.uk |
| Office of the Deputy Prime Minister | www.odpm.gov.uk |
| Secured Car Parks Scheme | www.theaa.com |
| Standing Committee on Structural Safety | www.scoss.org.uk |

# Appendix A
# Background of design, construction and structural performance

## A1 General

A1.1   There are over 4000 car park structures in the UK. Almost all have been built since 1940 and most are of reinforced concrete construction. Many of the earliest multi-storey car parks, constructed as early as 1920, were built as parking garages with valet parking. They were designed to low standards. These car parks later became self-park garages and eventually multi-storey car parks.

A1.2   Normal density concrete has been most commonly used for the construction of car park structures, but lightweight aggregate concrete and high alumina cement concrete may have been used in some cases. Many car park structures were built on the basis of competitive 'design and construct' contracts. Many were designed to the lowest first cost per parking bay with the optimistic assumption that Operators would provide appropriate care and maintenance. Over the years the standards of design and construction have improved following adverse experiences of performance of some early structures.

A1.3   Car park structures have generally been safe structurally but the performance record of the early bare utilitarian concrete structures has too often been marred by poor quality concrete and durability of the resulting structure. Although generally they had a notional design life of 50 years, many began to show considerable deterioration after a much shorter time, typically 15–20 years. Some structural failures have occurred.[A1]

A1.4   Compared with most forms of commercial building, car park structures have a number of distinct characteristics. The typical clear span is 16 m. This is a long span and the dead load/live load ratio is higher than for most forms of normal concrete building structure. In general a larger proportion of the live load is experienced for longer periods from the start.

A1.5   Internal floors are subject to wetting and drying. Decks flex under vehicle movement. The movement of the structural frame due to thermal and moisture changes requires consideration in design. This is especially important for top decks, which are often waterproofed with black materials, as they will experience the largest thermal movement. Large areas of structure are required and movement relief joints are unwelcome and if provided can fail to operate. Many have leaked or were wrongly positioned in early concrete car park structures.

A1.6 A shorter span between columns was acceptable in early concrete car parks (approximately 9 m) with cantilevers of approximately 3 m on either side. This was economical in the depth of the deck slab and the quantity of concrete used since the positive and negative moments in the parking deck were balanced. Many architects also had a preference for avoiding the appearance of columns on the façade. Cantilever performance is therefore critical in such structures. The penetration of de-icing salts into cracks in the top cover of the concrete can lead to a high corrosion rate and a loss of structural capacity in both cantilever and main span once concrete spalling occurs.

A1.7 Car park structures sometimes form part of a building with, for example, shops below and offices or flats above. This requires load transfer floors that change the column grid on non-parking floors, producing a heavy slab structure uncharacteristic of the other parking floors. Parts of such structures may therefore be subject to powerful movement differentials or stresses due to the different stiffness provided by the parking and non-car park parts of the building. The structure of many concrete car parks is also complicated by the complex geometry of ramp/circulation arrangements required to maximize access through the structure.

## A2 Structure

A2.1 During the 1960s, construction contracts were frequently based on specifications less demanding than is now known to be appropriate for achievement of durable car park structures. The structural codes CP114, CP115, CP110 and BS 449 were generally applied in the design but the proportions of car parks were unconventional compared with other building structures and their exposure conditions ambiguous. It was not until 1976 that the Institutions of Structural Engineers and of Highways and Transportation published the first authoritative guide to multi-storey car park design. The guide did not give comprehensive detailed advice on structural design, on waterproofing or on the deleterious effects of de-icing salt on reinforced concrete.

A2.2 Some early car park structures have been demolished prematurely. In several other cases they were strengthened. Such experiences resulted in changes to the design codes and to the IStructE/IHT guide, which was updated in 1984.[A2] However, this guide did not emphasize the severity of exposure of reinforced concrete associated with de-icing salt, freezing whilst wet, etc., and the need for high quality concrete to achieve durable car park structures. The guide did, however, draw attention to the risk of chloride-induced corrosion in concrete decks. Latterly concrete car park construction has been based on more up-to-date structural codes, although they usually treat car parks as a special case and do not classify the exposure conditions. The Institution of Structural Engineers has recently issued new design recommendations that give specific guidance.[A3] Car park structures designed and built following those recommendations are likely to be more durable and give better long-term structural performance.

A2.3 Premature corrosion of reinforced concrete decks and ramps has generally been due to:
- specification of a low strength grade of concrete, based on structural requirements alone without consideration of durability, in particular the effects of exposure to aggressive environmental conditions including de-icing salt;

- a low strength grade of concrete with a high water/cement ratio, resulting in highly permeable concrete with a very weak top layer owing to bleed effects;
- poor workmanship/compaction of concrete;
- lack of curing of the critical top surface, increasing the permeability of the concrete;
- lack of attention to construction tolerances, leading to low cover to the top reinforcing bars;
- surface cracking of the slab, leading to rapid chloride salt penetration direct to the reinforcing bars.

A2.4    In addition to corrosion of top reinforcement, corrosion of bottom reinforcement can be a problem, particularly in the soffit below the top deck slab and around leaking joints. Corrosion of bottom reinforcement can also be found in precast decks with insitu concrete topping, e.g. units where cover to bottom reinforcement may be inadequate (often less than 12 mm).

A2.5    Sometimes there is warning of corrosion before the structure becomes unsafe through the expansion of corrosion products resulting in cracking and spalling of the concrete. Such warning may not, however, be generated in the case of chloride-induced pitting corrosion in concrete deck slabs, which sometimes occurs locally, corroding reinforcing bars without any outward signs (see B5.1). Where previous repairs of a cosmetic nature have been undertaken to chloride-contaminated concrete, deterioration around the patches can be accelerated. From the structural safety point of view, the critical feature is likely to be undetected loss of cross-section of the bars at highly stressed locations, cracked critical areas (such as ramp to slab connections), or the combination of bending and high punching shear at columns supporting deck slabs.

A2.6    Where deck slabs are cantilevered, the tip of the cantilever sometimes supports heavy parapet walls, which may also serve as an edge beam acting as a load distributor along the full width of the cantilever. The redundancy inherent in such a system may enable substantial local strength losses to produce no serious immediate visible results. However, any consistent shortcomings in cantilever design or construction would be made more serious. Correct detailing and placing of reinforcement are critical.

A2.7    For early structures incorporating post-tensioned concrete components, the risks of tendon corrosion and structural failure without warning should be recognized as similar to pre-1992 post-tensioned bridge decks, although the risk is probably less and a longer time may elapse before safety is jeopardized. The continued integrity of the prestressing, whether by bonded or unbonded tendons, is dependent upon the ability of deck waterproofing (if present), the concrete and, for bonded tendons, the grouting to resist chloride migration. Chlorides are known to migrate even through good structural concrete. There is a risk of corrosion where de-icing salt is carried by vehicles onto decks from the highway, or spread on them by hand. This risk may be of concern in the early life of a car park structure owing to ingress at anchorages and joints. Recent surveys of new car park structures have shown chloride ingress occurring into concrete decks in the first 2–5 years, indicating that significant corrosion damage is likely in the medium term.

A2.8    Column/deck slab connections and column head/beam zones are perhaps the most critically stressed areas and most difficult to detail effectively to avoid cracking. In early concrete flat slab structures, reinforcing bars running through the columns were used in both directions. Lift-Slabs, however, had no bars running through the columns, making the detailing of the bars around columns critical. In these structures there is sometimes no bottom steel, and poor detailing and curtailment of reinforcing bars in the vicinity of the supports.

A2.9    Code rules changed with regard to the shear provision around column heads in 1972 and 1984. Also in 1972, amendment was made to encourage provision of more negative moment reinforcement at internal supports. Current codes do not require bottom reinforcement in concrete slabs directly through columns. They do, however, require 25% of bottom reinforcement to be properly anchored at the support. Ductile behaviour in punching shear may not therefore be ensured if detailing is poor. Punching failure may be sudden, especially if the slab is deteriorated, giving rise to increased risk of progressive collapse.

A2.10    Failure of a column and redistribution of load to adjoining columns following vehicle impact is another possible cause of disproportionate collapse of a car park structure. The resistance of early structures to progressive collapse following local damage or failure, i.e. robustness, was often not explicitly considered in design. Some structures, particularly those built of precast components, may have low robustness. Consideration of resistance to normal loads and to accidental loading is necessary in the Structural Appraisal of an existing concrete structure where there is a likelihood of impact from errant vehicles. The structure should be resistant to local damage, which should not propagate to a disproportionate extent.

A2.11    Thermal and moisture changes may cause significant movements and result in cracking distress if movement is restrained, e.g. by a lift shaft. These changes are not generally critical in most car park structures as spans are long and columns are slender so that there is little carry over of moment. However, expansion/contraction phenomena have caused structural distress in some insitu concrete structures. The most serious consequences have generally been cracks in columns. Warnings of structural distress would generally be evident well before serious consequences are likely. Thermal effects may, however, increase susceptibility to sudden punching shear failure in deck slabs, especially where steel corrosion and concrete deterioration or concrete repair is also present. Susceptibility of such structurally sensitive areas may also be increased if the concrete is weak, porous and unprotected by a membrane, and freeze/thaw action causes degradation of slabs, especially top deck slabs.

A2.12    Totally precast forms of car park construction may include wall-type columns in combination with deck units of double tee or I-beams similar to those used in bridges. Compared with insitu reinforced concrete construction, precast concrete manufacture generally uses higher strength concrete and better control of reinforcement cover resulting in more durable concrete components. Precast concrete systems involve stress concentrations at the deck bearings, the condition of which should be monitored by inspections during service. Totally precast concrete structures generally exhibit some long-term movement and cracking as the precast units 'rock' on their seatings. Movement at supports makes effective sealing to prevent water leakage difficult to achieve. Cracking of columns has been observed in recent structures built using very long precast prestressed units.

## A3   Cladding

A3.1    For the safety of users, ventilation of multi-storey car parks is essential. For this purpose, the parking decks of multi-storey car parks are usually open to the weather and the use of external cladding is limited to protecting the external faces of deck edges and the structure of access stairs and lift shafts from the weather. The cladding on deck edges may also be extended upwards to provide edge protection for pedestrians usually with, in addition, a metal balustrade on top. Alternatively, an insitu reinforced concrete upstand may have been built at the deck edge to provide edge protection and to which external cladding is fixed. Masonry-faced concrete upstands are common and may lack ties, or have corroded ties. In some cases cladding, open in form to allow ventilation through it, may extend between parking decks and entry/exit ramp structures to improve the visual appearance of the structure. Cladding components are generally of precast reinforced concrete or masonry. Less commonly, metal cladding components are found and, in more recently constructed multi-storey car parks, metal and glass components have been used to provide enclosures for pedestrian stairs and lifts.

A3.2 Broadly, inadequacies found in concrete or masonry claddings may be considered in categories: the components are deteriorating, and/or they are out of position, and/or they are inadequately designed to accommodate movement or to remain fixed to the structure. Cracks, spalling, bulging and defects due to movement that are not visible may arise from many causes. It is essential to determine the causes before any repair or remedial action is considered.

A3.3 Precast reinforced concrete cladding commonly deteriorates as a result of the corrosion of reinforcement arising from the presence of chloride ions, carbonation and/or poor quality manufacture, in particular low concrete quality and cover. Concrete cracking and spalling will usually occur as deterioration progresses. The distress shown may be exacerbated by movement of the components due to dimensional changes, movement of the supporting structure or, more likely, inadequacy of fixings. Fixings are generally metal and inadequacy may arise because the number and quality necessary for safety were not installed, they were not properly installed during construction and/or they have deteriorated in use.

A3.4 Masonry cladding itself is generally quite durable but it may exhibit distress due to dimensional changes, movement of the supporting structure and/or corrosion of wall ties. The safety of the cladding can be in doubt owing to lack of wall ties and masonry panel support, poor workmanship during installation and/or inadequate design. These shortcomings cannot usually be seen during a visual inspection.

A3.5 Design requirements and recommendations for concrete and masonry claddings of buildings have improved over the years.[A4] Cladding is therefore likely to be inadequate and in poorer condition on early car park structures, given also the longer time of exposure to the environment. Where cladding may be at risk from vehicle impact, it should have been designed with this possibility in mind or protected to minimize the risk of impact damage.

## A4 Edge protection

A4.1 Barriers around the edges of the decks of car park structures should protect pedestrians. They should also reduce the risk of vehicles being driven, inadvertently or otherwise, over the edge of the deck. Barriers should not deflect on vehicle impact to such an extent that they impact into exterior cladding, damaging it or causing it to break free and fall to the ground below.[A3] Incidents of these types have been reported.[A1]

A4.2 Accidents involving children falling from car park deck barriers have not come to notice but potential accidents have been reported. Some car park barriers provide an irresistible temptation to young children who can easily climb them and so be placed at risk. Whilst parental watchfulness is the main safeguard, it is clearly desirable that barriers should be designed as far as is reasonably possible to prevent children easily climbing on them. For children under 5 years old, this requirement can be met by making the height of the handrail 1100 mm above the level of the highest foothold (see A4.12).

A4.3    Several accidents have occurred in which a car was driven through an edge barrier and plunged to the ground below. The existing edge barriers and/or their fixings were insufficient to restrain the car in each case.

A4.4    Fixings can be inadequate and can become more so as they corrode. The extent of corrosion may not be apparent, especially where the fixing is encased or attached to concrete. Load testing may then be necessary to determine adequacy. Some forms of metal edge barrier are susceptible to brittle failure on impact and need to be replaced or strengthened to provide a ductile mode of failure at a higher load.

A4.5    No specific design guidance on the provision of adequate and consistent edge protection was available to designers prior to 1972. Consequently about one-quarter of all multi-storey car parks in the UK were constructed without reference to a national design standard for edge protection. As a result, many early edge protection systems were inadequate for the purpose and were also subject to loss of structural capacity due to deterioration in use.

A4.6    A standard for the impact loading on an edge barrier due to a road vehicle was first published in 1972 as a supplement to British Standard Code of Practice CP3: Chapter V: 1967.[A5] The requirement for the edge barrier was that it effectively contained a 1500 kg car, impacting normal to the barrier at 10 mph. A method for calculating the equivalent static force to be resisted was given.

A4.7    The current British Standards are BS 6399: Part 1[A6] and BS 6180.[A7] They include the same basic requirement for perimeter vehicle restraints in multi-storey car parks. For car parks designed to carry vehicles of up to 2500 kg gross mass, the perimeter restraints are required by these standards to contain an impact from a 1500 kg car travelling at 16 kph (approximately 10 mph) impacting normal to the line of the barrier. For design, this force is distributed over any 1.5 m length of barrier at a height of 375 mm above deck floor level.

A4.8    For vehicle impact on bridge and other columns adjacent to highways, requirements for resistance to impact forces are given in the *Design manual for roads and bridges*,[A8] in particular BD 37/88, BD 52/93 and BD 60/94. British Standards BS 5400[A9] and BS 6779: Part 1: 1998[A10] are also relevant. Eurocode 1: Part 2.7: *Accidental actions due to impact and explosions*[A11] gives requirements that are expected to become a European Standard.

A4.9    National standards for barrier impact resistance have been established in other European countries and in the USA.[A12]

A4.10   A recently completed study of edge protection in multi-storey car parks[A12] has proposed some changes to the specified design requirements for vehicle edge restraints in BS 6399: Part 1 to take into account changes in the characteristics of vehicles. For simple situations at deck edges, the adoption of an impact height of 445 mm (instead of 375 mm) and full-scale compliance testing are proposed for edge protection systems. No changes are proposed, however, to the vertical design load for vehicles (2500 kg), the 1500 kg mass at 16 kph horizontal impact loading, the 1.5 m impact width or the assumed vehicle impact deformation (100 mm).

A4.11 For edges to decks in some particular situations, i.e. on ramps, by stair wells, at the end of long access lanes, and at edges of split level decks, some changes to BS 6399 requirements are also proposed. Greater stair well edge protection is proposed (increased to the same level as for ramp edges). For edges where there is a straight vehicle approach length exceeding 20 m on decks without traffic calming measures, the proposed static design force to be resisted by the edge barrier is twice that generally used at 445 mm height. The proposed provision at split-level deck edges is the same as generally applied but with a relaxation of deflection criteria where designated pedestrian routes do not pass adjacent to and beneath the barrier.

A4.12 For the protection of pedestrians against falling from deck edges, recent changes have been made to the requirements given in the Building Regulations, BS 6399: Part 1[A6] and BS 6180.[A7] Pedestrian edge protection should be provided to a height of 1.1 m measured above the highest foothold. In general edge protection is made more difficult to climb by eliminating footholds in design.

A4.13 The changes to requirements described above have been proposed by the Steering Group for the DETR study.[A12] They may be adopted formally in British and European Standards in due course. It is recommended that they are considered when specifying replacement or upgrading of edge protection systems in multi-storey car parks. An assessment method for installed restraint systems was defined also in the DETR study.[A13]

## A5    References to Appendix A

A1. *Structural Safety 1994–96: Review and recommendations.* Standing Committee on Structural Safety, 11th Report, January 1997.

A2. *Design recommendations for multi-storey and underground car parks.* 2nd edn. Institution of Structural Engineers and Institution of Highways and Transportation, London, 1984.

A3. *Design recommendations for multi-storey and underground car parks.* 3rd edn. Institution of Structural Engineers, London, 2002.

A4. *Aspects of cladding.* Institution of Structural Engineers, London, August 1995.

A5. *Code of Practice CP3: Chapter V 1967: Loading: Part 1: Dead and imposed loads,* British Standards Institution, London, 1967, Supplement 1972.

A6. *BS 6399: Part 1: Code of practice for dead and imposed loads.* British Standards Institution, London, 1996.

A7. *BS 6180: Code of practice for barriers in and about buildings.* British Standards Institution, London, 1995.

A8. *Design manual for roads and bridges,* Highways Agency, London.

A9. *BS 5400: Steel, concrete and composite bridges.* British Standards Institution, London.

A10. *BS 6779: Part 1: Specification for vehicle containment parapets of metal construction.* British Standards Institution, London, 1998.

A11. *EN 1991-1-7: Accidental actions due to impact and explosions,* CEN (in preparation).

A12. *Edge protection in multi-storey car parks – design, specification and compliance testing.* DETR Partners in Innovation Contract 39/3/570 cc 1806. Report, October 2001.

A13. *Edge protection in multi-storey car parks – assessment method for installed restraint systems.* DETR Partners in Innovation Contract 39/3/570 cc 1806. Report, October 2001.

# Appendix B
# Defects, cracking and deterioration processes in structural concrete

## B1 General

B1.1 Structural defects and inadequacies that may be found in car park structures are described briefly below. Non-structural cracking of reinforced and prestressed concrete in car park structures, cladding or edge barriers, i.e. cracking that is not due to the loads applied to the structure, may arise from deficiencies in detailing and fixing of reinforcement, poor quality in the original construction and/or deterioration of the concrete or corrosion of reinforcement over time. In general poor original concrete quality increases the susceptibility of concrete to deterioration processes.

B1.2 There is substantial literature elsewhere on structural defects, cracking and deterioration processes of structural concrete, which should be referred to for more comprehensive information.[B1–B4]

## B2 Cracking due to deterioration processes

### B2.1 Cracking of concrete due to reinforcement corrosion

B2.1.1 Damage to concrete structures by reinforcement corrosion is described comprehensively elsewhere.[B2–B4] In brief, embedded steel reinforcement and prestressing tendons are normally chemically protected from corrosion by the alkalinity of the concrete. In car park structures, the protection is likely to be lost because chlorides are present in uncarbonated concrete through the use of chloride-contaminated aggregates, calcium chloride during concrete mixing, or contamination by de-icing salt whilst the car park is in use. Chloride contamination generally occurs through de-icing salts imported into the car park on vehicles or manually spread onto car park decks during the winter months. For multi-storey car parks that are only a short distance from the sea, chloride contamination of the structure, cladding and edge barriers can also arise from wind-blown sea spray.

B2.1.2   Where concrete is contaminated with chlorides, the corrosion mechanism can lead to a very localized severe loss of steel cross-section (pitting corrosion). Expansive pressures may not build up within the concrete causing the concrete cover to crack and spall off (as usually happens when corrosion occurs in carbonated concrete; see B2.1.5). There may therefore be no external visible signs on the concrete surface of pitting corrosion, except that sometimes corrosion products may seep out and produce brown stains on the concrete surface. Brown stains can arise from other causes, e.g. iron pyrites in aggregates, but they should always be investigated since pitting corrosion can lead to substantial loss of structural capacity of the component without any signs of structural distress becoming apparent.[B2]

B2.1.3   Once reinforcement has become active owing to the effects of chloride ions, the corrosion process is driven by galvanic action, forming anode and cathode sites along the bar and leading to pitting corrosion. Although the corrosion pits can be as small as a few millimetres in diameter, the galvanic corrosion cells can produce a rapid loss in cross-section of the reinforcement in highly localized areas along the reinforcing bars. Such pitting corrosion is characteristic of all chloride-induced reinforcement corrosion.

B2.1.4   Another important characteristic of galvanic corrosion is that it can occur under conditions that are normally unfavourable to corrosion, such as where concrete is in permanently wet conditions, which would be expected to reduce the oxygen supply rate to the corrosion and slow the process down. However, when driven by galvanic action, the steel at the anode is converted into soluble ferric chloride; this leads to a reduction in the cross-section of the reinforcement without the bursting pressures that produce visible signs at the concrete surface.[B5]

B2.1.5   The protection of reinforcing steel from corrosion can also be lost if the concrete becomes carbonated by carbon dioxide in the atmosphere. In such cases, an expansive corrosion product is usually produced that generates bursting stresses in the concrete leading to cracking of the concrete cover. Some rust staining of the concrete surface may also appear. Eventually the concrete cover will spall off as the corrosion product continues to build up. The corrosion tends to progress uniformly over a large area of reinforcement bar. The time elapsed before corrosion of embedded steel commences and before cracking and spalling of concrete takes place depends on the concrete quality and cover to the steel and on the exposure conditions.

## B2.2   Alkali silica reaction

B2.2.1   Concrete made with aggregates containing reactive silica can be subject to alkali silica reaction (ASR) when sufficient alkali (usually from the cement) and water are present. The reaction produces a hygroscopic gel within the concrete generating internal expansive forces. In plain or reinforced concrete, distinctive map cracking of the concrete may then develop in the form of a regular network of interconnected cracks and the gel may emanate from the cracks. In prestressed concrete, cracking may not be in the distinctive map cracking form.

B2.2.2   ASR does not usually produce the distinctive crack pattern in concrete until the structure is more than 5 years old. The presence of ASR can, in some cases, substantially impair the long-term durability of a concrete element. Once suspected from observation of cracks and their pattern, a Structural Investigation is needed, including laboratory testing of specimens, to confirm whether ASR is present.[B6] Guidance on assessing concrete structures affected by ASR is given elsewhere.[B7]

### B2.3  Freeze/thaw damage

B2.3.1  Concrete situated in a wet environment and subjected to freeze/thaw cycles can be disrupted if it is of inadequate quality. Water in the pores of the concrete expands on freezing, causing internal stresses that can fracture the near-surface concrete. In particular, top decks of car park structures are frequently saturated, giving ideal conditions for freeze/thaw action, leading to surface scaling and reduced mechanical properties of the concrete. Guidance on assessing concrete structures affected by freeze/thaw action is given elsewhere.[B7]

### B2.4  Failure of concrete repairs and protection systems

B2.4.1  Repairs to cracked or spalled concrete may fail over time. Fresh cracking of the repaired areas and delamination of the repair material may occur as corrosion continues within the repair owing to inappropriate repair or wrong diagnosis of the causes of the corrosion. Areas around the original repair may also spall if chloride-contaminated concrete was not removed before the repair was made. Corrosion continuing within the repair is not common. A more likely cause of failure is the effect of inadequate bond of the repair through poor surface preparation and treatment or inadequate concrete breakout such that steel reinforcement is not embedded in the new repair material. Failure may also result, as mentioned in 15.3.5, from incompatibility between the repair material and parent substrate (see 10.6).

### B2.5  Sulphate attack

B2.5.1  Sulphate attack has been found in some car parks arising from a misuse of materials containing gypsum placed on the decks to mop up oil droppings and remove stains. When wet, ettringite is formed in the concrete pores and expands and causes spalling/shelling of the surface, which can be harmful to finishes. When dry, the ettringite appears on the surface as a white powder. In severe cases of sulphate attack, exposed aggregate can work loose, exacerbating erosion of the running surface.

B2.5.2  Sulphate attack can also exist in buried concrete in contact with sulphate contaminated soils and groundwater. Recent investigation of buried concrete structures in the UK has revealed a form of sulphate attack, known as thaumasite attack, caused by specific environmental conditions. Guidance on action relating to this form of attack is available elsewhere.[B5]

## B3  Structural defects and inadequacies

### B3.1  General

B3.1.1  Beams, deck slabs, columns and cladding may exhibit signs of structural distress due to one or more of the following:

- inadequate design;
- errors in detailing and fixing of reinforcement;

- the use of materials of low quality in construction and/or poor workmanship;
- early stripping of forms and ineffective back-propping;
- in the case of Lift-Slab construction, unsynchronized jacking and lifting;
- significant deterioration of the structural materials;
- loading in excess of design loads;
- damage due to vehicle traffic, impact or fire;
- abrasion due to wear and tear causing removal of cover to the reinforcement in the tops of beams in areas where it is required to resist support moments;
- unforeseen movement.

## B3.2 Structural cracks in concrete

B3.2.1 There are many causes of structural cracks in concrete. Cracking of concrete in shear, torsion or flexure may indicate structural defect or inadequacy. Shear cracking is usually only visible in beams and slabs. Typically cracks propagate at an angle of about 45° from the junction of the beam or slab with its support. The cracks tend to be widest at the support and to be few in number. Monitoring of the width and length of shear cracks can provide a useful indication of whether they are becoming more severe over time. Torsion cracking of concrete beams or slabs may have a similar form to shear cracks. Damage to deck slabs can be caused by deflection of edge barriers following a vehicle impact.

B3.2.2 Flexural cracking in simple-supported concrete components usually appears in the middle third of the span. Flexural cracking is usually in the form of several parallel cracks running at right angles to the direction of the span. In continuous construction, cracks may not be within the middle-third zone. In these cases, and also for two-way spanning slabs, the Engineer may establish the cause of observed cracking through determining the load paths through the structural component and its response. The monitoring of cracks over time will usually assist the Engineer in determining their cause and whether their propagation is progressive.

B3.2.3 Local distress of a concrete beam, column, deck or ramp manifest as shear, torsion or flexural cracking may indicate that the structure is unsafe. Such cracks may also be caused by weakening of the concrete or corrosion of embedded steel as a result of deterioration in the materials (see B4). However, deterioration processes may cause no apparent distress or the degree of deterioration may not be visibly great but the component may have an inadequate reduced capacity. Some types of deterioration, such as pitting corrosion of reinforcement or chemical degradation of concrete, may not be manifest at all visibly but may have reduced the load-carrying capacity of the component to a critical level (see B2.1).

B3.2.4 If cracking is the result of structural distress and is detrimental to the stability of the structure, the structure should be propped immediately, and before undertaking testing and/or a detailed assessment of the likely cause of the cracking.

## B3.3 Impact damage

B3.3.1 Impact damage to car park structures is usually caused by accidental or careless use of motor vehicles. Impact damage due to vehicles may also arise at locations on decks and ramps where the surface is uneven, e.g. at expansion joints. The damage to the structure is usually obvious. Most often edge barriers are damaged but damage to columns, brickwork and cladding does occur where they are unprotected or the protection of barriers is breached. A very severe impact may cause movement of structural components. In such cases a Special Inspection may be needed to ascertain the true extent of the structural damage.

## B3.4 Fire damage

B3.4.1 Fires due to burning vehicles or goods in multi-storey car parks, if severe enough, may damage the structure and require remedial works.[B8] Concrete and masonry are, however, quite resistant to the effects of fire. Structural steel can become weakened substantially at temperatures that may arise in vehicle fires.

B3.4.2 The visible evidence of fire damage may be limited to black smoke deposits around the edge of the fire location. However, heat from the fire may have affected the strength and durability of the concrete or even the yield strength of the reinforcing bars or prestressing.

B3.4.3 For moderately severe fires, heat expansion of the concrete may cause explosive spalling of aggregate in the concrete cover. In severe cases, spalling of the concrete may occur owing the large temperature differentials in the concrete. Such spalling can occur at relatively low temperatures and may leave the remaining concrete unaffected. However, once the temperature of the concrete reaches 600°C, concrete strength can reduce substantially. A Special Inspection would be needed following a severe fire within a multi-storey car park.

## B3.5 Movement of ground and/or structure

B3.5.1 Ground movement or movement of the structure or of individual components, e.g. at construction joints, expansion joints or cladding, is usually visible. Cracking, tilting or distortion of the primary structure or excessive movement at joints may be due to settlement of foundations or ground movement. Possible causes of ground movement include freeze/thaw action, changes in the level of the water table and mining subsidence. The observed effects of the movement often enable the cause to be postulated. More detailed inspection or investigation can then be targeted to confirm the cause.

B3.5.2 Car park structures below ground are generally of reinforced concrete construction. They may be subject to similar movement defects, buoyancy effects and/or deterioration as basements of buildings more generally.[B9]

## B3.6 Cladding

B3.6.1 Cladding on car park structures is usually made up of concrete or masonry panels. Structural deficiencies are most commonly due to inadequate fixings or supports, either as originally constructed or through deterioration over time of the fixings themselves or of their anchorages in the structure or cladding panel.[B10]

B3.6.2 Cladding may also be damaged by vehicle impact or by movement of the supporting structure, which may arise from a wide variety of causes.[B10] Concrete cladding materials may also deteriorate by the same processes as for concrete more generally (see B2). Concrete cladding and fixings are likely to be most severely deteriorated at locations where they are exposed to de-icing salt. Masonry panels may be damaged by frost action causing spalling of the faces of bricks.[B7]

## B3.7  Edge protection barriers

B3.7.1  Edge protection barriers may deteriorate over time and/or they may be damaged by vehicle impact. Deterioration most commonly occurs owing to corrosion of metal components or fixings. The deterioration is generally visible except in fixings where bolts and other parts of connections are hidden from view within connections or encasing concrete. At these locations, concrete is likely to be damp, contaminated by chlorides or carbonated, providing conditions for accelerating corrosion.

# B4  Materials quality and deterioration

## B4.1  Loss of concrete

B4.1.1  Loss of concrete from car park structural components may arise from accidental impact or fire damage or, in decks and ramps, from the effects of vehicles running over an uneven surface, e.g. at expansion joints. More commonly, however, loss of concrete is caused by internal expansive forces induced by deterioration processes, in particular corrosion of embedded steel, ASR or freeze/thaw action, that fracture and delaminate the surface layer of concrete. Local overstress of a concrete component may also result in loss of concrete.

B4.1.2  Delamination is the detachment or partial detachment of an area of concrete from the parent mass. There may be little visible sign of the defect as the edges of the delaminated area remain attached and the surface of the concrete may be only slightly lifted. As the deterioration progresses, the delamination may be indicated by cracking on the concrete surface and perhaps by rust staining. Eventually the delaminated concrete may break free, becoming fully detached and falling away, leaving an area of spalled concrete. It should be noted that delamination may not occur where corrosion is chloride induced and in the form of local deep pitting of reinforcing bars.

B4.1.3  Delamination and spalling degrade the appearance of the structure and can cause a rapid acceleration of deterioration, especially through corrosion of exposed steel reinforcement. In critical locations or severe cases, such deterioration can impair the load-carrying capacity of the structure. Measurement of the extent of deterioration, and in particular loss of steel cross-section, is required to enable the current adequacy and safety of the structure to be appraised. Determination of the weakening of the structure due to pitting corrosion, which may occur in chloride contaminated concrete, is particularly difficult because this form of corrosion occurs very locally.

B4.1.4  It should be noted that loose, but attached, delaminated concrete poses a significant risk of debris falling onto any people below. This situation can arise through neglect of the structure. If found, such loose concrete should be removed as soon as possible after it has been identified in order to restore safety. The relevance of the spalled concrete to the stability of the structure should be checked and, if necessary, the concrete should be made good structurally with effective concrete repairs.

### B4.2 Cracks in concrete

B4.2.1 Cracks in concrete structure, cladding or barrier components may arise from many causes. They may be passive, i.e. they do not open or close owing to temperature change or loads on the structure, and affect appearance only, e.g. plastic settlement cracks formed when the concrete was cast. Alternatively, cracks may be active, changing in width as the structure flexes and progressively widening and lengthening over time. Active cracks may be indicative of an underlying structural defect. Movement of newly observed cracks should be monitored. The orientation and depth of cracking may considerably influence the durability of reinforcement.[B4]

B4.2.2 Cracks before the concrete has hardened can be caused by plastic shrinkage, plastic settlement, formwork movement during casting or poor workmanship during placing. In hardened concrete, the possible causes of cracking include long-term drying shrinkage or crazing, reinforcement corrosion, ASR, carbonation or thermal action. Propensity to cracking can be increased by inadequate curing of concrete in all weather conditions. Cracking has also been experienced in concrete in thin sections where Portland cement replacements were used in the concrete mix, giving low early age tensile strength.

B4.2.3 In some cases, cracking is caused by a progressive deterioration process that, over time, may impair the structural safety of the construction. The different types of non-structural cracking arising from poor quality construction and/or materials deterioration processes are described briefly in B1 and B5, together with their causes and significance.

## B5 Cracks arising from construction processes

### B5.1 Plastic cracking

B5.1.1 Plastic cracks that can appear shortly after concrete is placed may be either plastic settlement cracks or plastic shrinkage cracks. Settlement cracks are a common feature of car park deck slabs, usually over beam strips.[B1] Hogging over stirrups is found with cracking along the line of the reinforcing bars. Plastic settlement cracks may also be found in columns or walls, with the cracks forming along the line of horizontal steel reinforcement bars as the concrete slumps between them. Voids may be present under the bar. The concrete surface may have an undulating appearance. Rapid drying of the surface of freshly placed concrete causes plastic shrinkage cracks. They occur more commonly in concrete deck slabs. They do not usually follow the lines of the reinforcement and the pattern may be random polygonal. The cracks may be minor but can pass right through slabs. Very fine cracks do not have a detrimental effect in concrete in moderate exposure conditions, but in car park decks they provide paths for de-icing salt, water and carbon dioxide to penetrate the concrete.

### B5.2 Early thermal cracking

B5.2.1 Thick walls and slabs may exhibit early thermal cracking. It is unlikely to be found in car park decks, as they are usually relatively thin. The cracking is a consequence of the heat of hydration of the cement in the concrete mix. A thermal gradient develops, with the outer part of the concrete cooling more rapidly than the core. The cooling surface concrete contracts and is restrained by the core. Cracking may then occur at the surface. Propensity to this form of cracking is reduced where pulverized fuel ash or ground granulated blast furnace slag has been used in the concrete mix. Similar early thermal cracking can also occur where concrete is cast onto a previously constructed concrete component that restrains the contraction of the setting concrete.

## B5.3 Drying shrinkage cracks

B5.3.1 Concrete tends to shrink as it loses moisture over several months or years. This long-term drying shrinkage can give rise to cracks if the concrete is restrained, e.g. by reinforcement. Drying shrinkage cracks can also arise soon after construction if the concrete is not cured correctly.

B5.3.2 Shrinkage of hardened concrete is related primarily to the aggregates used in the concrete. Some aggregates, i.e. shrinkable aggregates and lightweight aggregates, can produce concretes with high drying shrinkage. The random crack pattern caused by shrinkable aggregate may be similar to that caused by ASR. In some cases, carbonation of the concrete over time can cause shrinkage, usually known as carbonation shrinkage. It is likely to be a minor issue in most car park structures.

B5.3.3 Drying shrinkage cracks are likely to occur in thin slabs or long walls. They are usually very fine in width as shrinkage should have been taken into account in design, but they may widen over time in combination with other effects. Drying shrinkage cracks in themselves are generally of little consequence to structural integrity.

## B5.4 Crazing

B5.4.1 Crazing is the term used to describe an irregular closely spaced and small-scale map pattern of very fine cracks on the concrete surface. Crazing can occur in vertical faces against the face of formwork or on the upper surface of a slab. It is usually caused by shrinkage of the surface relative to the concrete mass as moisture movement occurs during curing of the concrete. The very fine cracks do not affect the structural integrity of the concrete and have little influence on deterioration of the surface, but may increase the initial carbonation rate.

## B5.5 Cracking due to poor compaction

B5.5.1 Delay in the concrete pouring sequence of slabs can result in surface cracks and honeycombing of the concrete in the interface region where fresh concrete was placed adjacent to previous batches that had already stiffened or hardened. Cracks and honeycombing of concrete in slabs, beams and columns may also arise owing to inadequate workmanship and poor detailing of the design. The cracking and honey-combing may structurally impair the component. The extent of such defects should be carefully investigated. Cracks may also arise in concrete repairs if poorly executed or inappropriate materials are used.

## B5.6 Cracking due to formwork movement or ineffective propping

B5.6.1 Where formwork moves during construction before stiffening of the placed concrete, a step is likely to form on the surface of the concrete, which may also be poorly compacted in the vicinity of the step. Honeycombing may also be caused following grout leakage.

B5.6.2 The failure of falsework support systems or their premature removal during construction can also result in cracking defects in the concrete. In extreme cases the load-carrying capacity of the concrete structure can be impaired. Unsynchronized jacking and lifting during construction of Lift-Slab structures can also cause cracking.

B5.6.3 Cracking can be caused by ineffective propping, during breaking out and repairs, which causes moment transfers and excessive deflections.[B11]

B5.6.4 Irregularities in the surface profile of slabs caused by formwork movement during construction are not usually structurally significant. However, honeycombing and cracks, which may also arise, may impair the long-term durability and strength of the structure.

# B6 References to Appendix B

B1. *Non-structural cracks in concrete.* Concrete Society Technical Report No. 22, 3rd edn, 1992.

B2. Currie, R. J. and Robery, P. C. *Repair and maintenance of reinforced concrete.* BRE Report BR254, 1994.

B3. *Corrosion in steel and concrete. Part 1: Durability of reinforced concrete structures. Part 2: Investigation and assessment. Part 3: Protection and remediation.* BRE Digest 444, February 2000.

B4. *The relevance of cracking in concrete to corrosion of reinforcement.* Concrete Society Technical Report TR 44, 1995.

B5. *The thaumasite form of sulfate attack: risks, diagnosis, remedial works and guidance on new construction.* Thaumasite Expert Group, Report, January 1999.

B6. *Structural effects of alkali-silica reaction.* Institution of Structural Engineers, London, July 1992.

B7. CONTECVET: *a validated user's manual for assessing the residual service life of concrete structures: Manual for assessing concrete structures affected by ASR; Manual for assessing concrete structures affected by frost; Manual for assessing corrosion-affected concrete structures.* EC Innovation Programme, IN309021. British Cement Association, 2002.

B8. *Assessment and repair of fire-damaged concrete structures.* Concrete Society Technical Report TR33, 1990.

B9. *Design and construction of deep basements.* Institution of Structural Engineers, London, August 1975.

B10. *Aspects of cladding.* Institution of Structural Engineers, London, August 1995.

B11. Canisius, T. D. G. and Waleed, N. *The behaviour of concrete repair patches under propped and unpropped conditions – critical review of current knowledge and practice.* FBE Report 3, March 2002.

# Appendix C
# Testing and monitoring concrete and steel materials and components

## C1 Introduction

C1.1 The more common test techniques used in Structural Investigations of the construction materials and components in concrete structures are briefly described below. More comprehensive information on available techniques and on the planning and procedures of testing and monitoring existing concrete structures may be found elsewhere.[C1-C8] The development of test techniques and standards is continuing and current British and European Standards should be consulted for details of procedures. Specialist advice should be sought on recent improvements and innovations in techniques that may be relevant in any particular case.

C1.2 An effective Structural Investigation of a car park structure can be made only if an appropriate test technique is selected and applied at sufficient locations to enable an estimate to be made of the parameter of interest. Guidance on the reliability and precision of different techniques and on sampling rates is given elsewhere.[C1-C8]

## C2 Investigation of the dimensions and quality of reinforced and prestressed concrete

### C2.1 Cover meter

C2.1.1 This electromagnetic equipment is used to estimate the depth to steel reinforcement in concrete and to determine its orientation and distribution. Particular care is needed where the meter is used on lightweight concrete or concrete containing crushed rock aggregate incorporating ferrous or magnetic particles. Some pozzolans, especially fly ashes, contain magnetic particles and some sands contain particles of magnetite. Although the equipment is straightforward to use, considerable care and skill are required to obtain acceptable results. It is essential to confirm the calibration of the equipment by breaking out the cover down to a sample of reinforcement or measuring at a spall. More modern cover meters overcome many of the problems encountered with the earlier designs of this equipment.

## C2.2 Physical exposure and delamination surveys

C2.2.1 The cover of concrete, masonry or non-structural finishes can be removed using hand tools or hand-operated power tools. This simple 'breaking out' technique enables hidden parts of structures to be exposed locally, e.g. reinforcement, structural steel and cladding fixings, for visual examination and measurement. Survey by hammer tapping or chain dragging (on the tops of decks) can be used to identify delaminated concrete cover.

C2.2.2 For reinforced concrete slabs and other components, exposure of the reinforcement at sample locations gives valuable confirmation of non-invasive surveys, e.g. cover, half-cell potential measurements, where these suggest that depassivated reinforcement is present.

C2.2.3 Breaking out should not be started before a check has been made that structural safety would not be jeopardized. Temporary propping may be required.

## C2.3 Ultrasonic pulse velocity in concrete

C2.3.1 The quality and uniformity of concrete can be assessed by measuring the velocity of ultrasonic pulses through it. Access is generally required to opposite faces of the component under test. The method may be used to determine the presence of voids, cracks or other imperfections in a component and to give comparative indications of variations in the strength of the concrete in different components or along a given component.

C2.3.2 The instrument does not give a precise estimation of strength without calibration. Calibration against core strength tests improves the accuracy of strength estimation. The technique requires precise measurement of the distance between the two ultrasonic transducers. Moisture variations can affect the results. It is suitable for use only by experienced persons. Skill is required in the analysis of results.

## C2.4 Core testing

C2.4.1 Core samples, normally of a minimum 100 mm nominal diameter, may be cut from concrete for measurement of the strength and density. They may also be used to indicate the distribution of materials in the concrete, the concrete quality (voids, honeycombing, etc.), and for petrographic, microscopy and associated studies. After strength testing, the crushed cores can be chemically analysed to determine mix proportions, the presence of admixtures, chlorides and contaminants, etc. Cores may also be used for measuring the shrinkage, expansion and absorption properties of the concrete, and the depth of carbonation.

C2.4.2 Coring should not normally cut reinforcement. A cover meter should be employed prior to coring to minimize the risk of unintentional damage to reinforcement and any prestressing strands in the structure. Coring can be used to determine cover, steel type and size, but only where the Engineer has agreed that it is acceptable to damage the reinforcement. Accurate determination of cover and steel type and size can also be determined by localized concrete break out around prelocated bars.

C2.4.3 Where standard cores cannot be obtained, for example in small beams, smaller cores may be taken. Small diameter cores (50 mm diameter or less) are a convenient means of obtaining samples for inspection and chemical analyses.

C2.4.4 Core cutting is normally a 'wet' process, although the flush water can sometimes be collected with little spillage.

C2.4.5 Remedial action to the requirements of BS EN 1504[16] is necessary to repair the damage to the structure.

## C2.5 'Pull-out' and 'pull-off' tests for concrete

C2.5.1 These tests seek to provide a measure for the compressive strength of concrete by inducing internal fracture within the material. This can be achieved in a number of ways, which are classified as either 'pull-out' tests where an anchor is inserted into the concrete, or 'pull-off' tests where a disk is bonded to the concrete surface. The tests cause some minor damage to the surface, which requires repair. Compressive strength is estimated from calibration charts. 'Pull-off' tests are also used to evaluate tensile strength and tensile bond strength for repairs, coatings and structural strengthening. Multiple tests are required to give a representative result.

## C2.6 Strength of steel

C2.6.1 Testing of steel by measuring the tensile load required to rupture a standard specimen is frequently used to obtain measurements of the ductility of the material and the tensile strength. Other properties that may be measured, include elastic limit, yield point, proof stress and modulus of elasticity. Test pieces of standard size (between 150 and 250 mm long) are usually required. Their cross-section may be circular, square, rectangular or, in special cases, of some other form. Test pieces should generally be machined to the dimensions given in Standards, but some sections, bars, tubes, etc., may in certain circumstances be tested without being machined.

## C3 Investigation of composition and conditions in concrete

### C3.1 Carbonation

C3.1.1 The internal alkaline environment within concrete and mortars affords corrosion protection to embedded metal. It is degraded by the penetration of acidic atmospheric gases. The most common gas is carbon dioxide. The resulting reduction in alkalinity, termed carbonation, can be shown by the use of an indicator solution sprayed onto a freshly broken concrete surface. Phenolphthalein solution identifies alkaline zones by a purple–red coloration.

C3.1.2 Where it is not possible to break off pieces of concrete or form a broken surface between two closely spaced drill holes, the depth of carbonation can be determined by an incremental drilling technique, the drilling dust remaining colourless when sprayed with phenolphthalein solution in the carbonated zone. Tests are easy and quick to perform. Well-trained operatives are needed. Where drillings are used, surface damage is minor.

C3.1.3   A more precise indication of the progression of the carbonation front into concrete can be given by thin-section microscopy techniques.

## C3.2   Chloride content

C3.2.1   Chloride content may be determined by chemical analysis of concrete samples in the laboratory and by more sophisticated techniques such as X-ray fluorescence spectrometry. Samples are usually taken in areas of high negative half-cell potential.

C3.2.2   There are several commercially available kits (HACH and QUANTAB methods) for field tests to determine the approximate chloride contents of incrementally drilled concrete and mortar samples. The tests are generally only of sufficient accuracy to indicate the presence and approximate level of significant chloride ion content. These indications should be backed up by a proportion of laboratory-based determinations combined with tests on control samples. The tests are generally straightforward and quick to perform (30 minutes). Staff with limited specialist experience can perform them. It is necessary to have some facilities for sample preparation.

C3.3.3   The sampling regime should be carefully designed for the particular structure and the nature of the envisaged contamination (i.e. introduced at the time of construction or subsequently by ingress).

C3.3.4   Where chloride is suspected to have been added during construction of a car park structure, it can usually be identified by sampling the concrete in areas away from possible contamination by de-icing salt, e.g. the walls of stairwells.

## C3.3   Half-cell potential and resistivity measurement

C3.3.1   The half-cell technique involves measuring the electrical potential of embedded reinforcing steel relative to a reference half-cell (silver/silver chloride or copper/copper sulphate) placed on the concrete surface.[C5–C8] It gives an indication of the risk of corrosion of the reinforcement. The results are plotted on potential contour maps enabling localized zones of corrosion risk to be identified. The values of electrical potential depend on the moisture content of the concrete. The method is therefore markedly susceptible to weather conditions and the time of year. It does not indicate whether corrosion is actually occurring or the degree of corrosion that has occurred. Care has to be exercised to ensure satisfactory electrical coupling with pore fluids within the concrete (see Figure 13.2).

C3.3.2   The electrolytic resistivity of concrete can be measured to assess resistance to corrosion currents. The moisture content of the concrete influences the measurements.

## C3.4   Linear polarization resistance

C3.4.1   The linear polarization resistance technique involves the measurement of electrical properties of the embedded reinforcement at a given point in time, which are directly proportional to the corrosion rate of the reinforcement at the time of measurement.[C9] The equipment is attached either to a sensor placed directly onto the concrete surface, or to fixed or embedded sensors, and the control equipment takes readings via a connection to the reinforcement, usually automatically.

C3.4.2 The results require specialist collection and interpretation, as they are susceptible to operator error and the circumstances of the reading locations. The technique provides a measurement of corrosion activity actually occurring and is therefore susceptible to temperature and weather conditions. Care has to be taken to ensure satisfactory electrical coupling between the de-mountable sensors and the concrete and between the reinforcement and pore fluids within the concrete.

## C3.5 Cement content and cement/aggregate ratio

C3.5.1 The determination of the cement content of hardened concrete requires the facilities of a chemical laboratory. The techniques used depend on whether or not the aggregate grading and content are to be established, as well as the type of cement, cement content and aggregate used. If, by chance, samples of the materials used to make the concrete are still available, the inherent errors in chemical analyses can be considerably reduced.

C3.5.2 Petrographic and thin-section techniques are also able to estimate mix proportions. These methods involve a microscopic visual examination of specially prepared specimens. They are relatively more expensive, but do produce a more detailed understanding of the concrete.

## C3.6 Type of cement

C3.6.1 Most concretes are made with Portland cement: ordinary, rapid hardening, sulphate resisting or low-heat. These cements are not normally distinguishable by standard chemical analyses of concrete. A complete chemical analysis of the fine material from a sample of concrete can be compared with typical analyses of various types of cement. A mineralogical examination may also be helpful. The use of scanning electron microscopy is a particularly powerful tool, being able to perform chemical identification over very small areas (such as an individual cement clinker particle).

C3.6.2 Portland blast-furnace cement and high alumina cement (HAC) may be distinguishable in concrete visually by the colour of the matrix. However, colour may also be affected by the aggregate used and by carbonation, and extreme care is necessary. HAC can usually be identified by a simple chemical test or by petrography.

## C3.7 Admixtures and contaminants

C3.7.1 Many varieties of admixture have been used in concrete construction. Calcium chloride is the most likely one to be considered in a Structural Appraisal. The amount of chloride in a concrete sample can be determined in the laboratory or a good estimate can be obtained by a field test with strip indicators (see C3.2). The presence of sulphate may also need to be considered. It can be determined readily by laboratory analysis of a sample (see 3.10).

C3.7.2 The determination of other admixtures and contaminants, e.g. organic admixtures, sugars and metals, is not simple and is not usually required for car park structures. They can be determined by laboratory techniques such as X-ray fluorescent spectroscopy, infrared absorption and scanning electron microscopy. The assessment of dosage (which is frequently the prime objective of the test) is dependent on a precise knowledge of the admixture used.

## C3.8 Type of aggregate

C3.8.1 The general types of aggregate used in concrete may be immediately apparent on visual inspection. Where there is doubt, a petrographic examination of a thin section will identify them and can also give information on hardness, porosity, permeability, specific gravity and thermal properties, as well as potentially deleterious substances.

## C3.9 Latent expansion

C3.9.1 To confirm suspected alkali aggregate reactivity, laboratory tests are generally conducted on concrete samples taken from cores. These tests establish long-term latent expansion and other characteristics.

## C3.10 Sulphate content

C3.10.1 Where attack by sulphates on concrete is suspected, visual signs may include cracks containing white or crystalline deposits and, in severe cases, heaving of the concrete.

C3.10.2 A test for sulphate content can be carried out using the method given in BS 1881: 124: 1988. The sulphate content is normally expressed by weight of cement. A method to determine cement content is also given in BS 1881: 124. If conclusive proof of sulphate attack is required, samples may be analysed by petrography or by X-ray diffraction.

## C3.11 Water and gas permeability

C3.11.1 These tests are designed to assess the permeability of concrete in the surface zone. The quality of the concrete in this zone is critical to durability. Laboratory tests on material samples provide the most reliable assessment of permeability.

C3.11.2 The Figg and the 'CLAM' water and air permeability tests are relatively simple tests, which can be used on site to evaluate insitu concrete. These methods have been developed to overcome difficulties with the initial surface absorption test (ISAT) method (see C3.12). Figg tests require drilled holes approximately 10 mm in diameter and 40 mm deep. For CLAM tests, a 50 mm internal diameter steel ring is bonded to the concrete to isolate the test area. Reported field experience is limited, and these tests are most suitable for comparative purposes.

## C3.12 Initial surface absorption

C3.12.1 The ISAT measures the surface absorption of concrete. Considerable care needs to be taken. The test does not damage the structure. It is most suitable for comparative purposes. The results are affected by the nature of the surface. It is normally a site test.

C3.12.2 This test is used to assess the durability of insitu concrete. Tests are very sensitive to concrete quality and correlate with observed weathering behaviour. A minimum dry period of 48 hours is required before tests on surfaces exposed to weather. Inherent variations in initial moisture condition of the concrete need to be taken into account when interpreting results.

## C3.13 Absorption

C3.13.1 Tests are made on small cores, 75 mm in diameter, cut from the concrete. Considerable care needs to be taken when performing these tests. Absorption limits for concrete at different ages are specified in some British Standards for precast concrete products.

## C3.14 Air entrainment

C3.14.1 Entrained air voids in concrete are generally spherical and are incorporated into concrete during production to give resistance to freeze/thaw damage. For effective protection of the cement paste from freezing and thawing cycles, air voids range in size from about 0.2 to 1 mm. The percentage of entrained air voids, their size and distribution may be established by microscopy point counting methods on prepared samples.

# C4 Investigation of voids within or between components

## C4.1 Endoscope and borescope

C4.1.1 These devices may be used for the inspection of the integrity of cavity wall ties or fixings of cladding panels.[C10–C12] They can also be used to inspect the condition of prestressing tendons in ungrouted or poorly grouted ducts in concrete beams and slabs. Access is gained either through the duct wall or (rarely) via the end anchorage blocks. Extreme care is required to avoid damaging the prestressing tendons. Holes are drilled into the ducts to provide access for the optic device. This technique may not be practical in car park structures where the ducts are small in size.

## C4.2 Detection of wall ties and cladding fixings

C4.2.1 Metal detectors depend on the interactions between a coil or coils carrying alternating voltage and conducting or ferromagnetic (or both) materials. The devices are similar to cover meters. Certain ingredients in masonry units, such as iron compounds in bricks, can cause erroneous results. Some experience of operation is desirable. They are often used in conjunction with endoscopes/borescopes.[C10,C11]

# C5 Testing of edge protection barriers

C5.1 Where edge protection barriers are found to be deteriorated due to corrosion, especially at holding-down bolts, investigation by uncovering or dismantling a sample of bolts usually enables the extent of corrosion in the sample to be measured. The loss of strength of the fixing can then usually be estimated and the adequacy of the barrier as a vehicle restraint determined. Alternatively, a pull-out test may be carried out, especially where holding-down bolts have worked loose.

C5.2 In some cases estimates of fixing and barrier strength cannot be made with sufficient reliability. Compliance testing of the barrier installation may then be required.

C5.3 Compliance testing provides a direct indication of the adequacy of the barrier system to restrain vehicles.[C13] The test should be applied at a selection of locations to a section of the existing installation. It is a pseudo-static load test of the resistance of the barrier to the static force for the specified impact. It aims to verify the quality of the barrier materials, the strength of the immediate supporting members and the strength of the fixings to the car park structure. Care should be exercised especially when testing barriers that are attached to precast units spanning parallel with the barriers, since there can be a danger of pushing the edge precast unit off the supporting beam (see 15.3).

C5.4 Edge protection systems that have been demonstrated to provide adequate resistance to vehicular impact forces are identified elsewhere.[C12] The referenced report also provides predicted deflection data and design bolt forces for common barrier types tested to the proposed new impact height of 445 mm. Replacement edge protection should be assessed against these criteria. A companion report[C13] provides the equivalent deflection and bolt force data for barriers installed to resist a 375 mm high impact. Appraisal of existing protection systems may be against these criteria.

## C6　References to Appendix C

C1. Bungey, J. H. and Millard, S. G. *Testing of concrete structures.* 3rd edn. Blackie Academic and Professional, 1996.

C2. *Diagnosis of deterioration in concrete structures – identification of defects, evaluation and development of remedial action.* Concrete Society Technical Report No. 54, 2000.

C3. Kay, T. *Assessment and renovation of concrete structures.* Longman, 1992.

C4. *Testing and monitoring the durability of concrete structures.* Concrete Bridge Development Group Guide No. 2, The Concrete Society, 2002.

C5. *Half-cell potential surveys of reinforced concrete structures.* Concrete Society Current Practice Sheet 120. *Concrete,* 2000, 34, No. 7, pp. 43–45.

C6. *Measuring concrete resistivity to assess corrosion rate.* Concrete Society Current Practice Sheet 128. *Concrete,* 2002, 36, No. 2, pp. 37–39.

C7. De Vekey, R. C. *Corrosion of steel wall ties – recognition and inspection.* IP13/90, BRE, Garston, 1990.

C8. *Corrosion in concrete.* Reports. Trend 2000 Limited, December 2001.

C9. *Diagnosis of deterioration in concrete structures – identification of defects, evaluation and development of remedial action.* Concrete Society Technical Report No. 54, 2000.

C10. Johnson, M. A. E. and Fifield, B. E. *Remote visual inspection of voids and cavities – practical experience on post-tensioned concrete bridges. Structural faults and repair 93,* Vol. 1, University of Edinburgh, 1993.

C11. *Corrosion of steel wall ties – history of occurrence, background and treatment.* IP12/90, BRE, Garston, 1990.

C12. *Edge protection in multi-storey car parks – design, specification and compliance testing.* DETR Partners in Innovation Contract 39/3/570 cc 1806. Report, October 2001.

C13. *Edge protection in multi-storey car parks – assessment method for installed restraint systems.* DETR Partners in Innovation Contract 39/3/570 cc 1806. Report, October 2001.

# Appendix D
# Assessment of structural safety risk and appraisal of concrete car park structures

## D1  General

D1.1  This appendix describes assessment of structural safety risk in general terms. It also briefly discusses appraisal of remaining capacity, structural sensitivities and vulnerability to progressive collapse of concrete car park structures. More detailed and comprehensive guidance may be found elsewhere (see Section 14).

## D2  Risks to structural safety

C2.1  The assessment of structural safety risk requires careful consideration and use of engineering judgement. The Engineer can determine relative risks qualitatively based on a consideration of the factors that influence them. Lack of knowledge precludes full quantitative risk assessment. The task of assessment is an essential part of the Structural Appraisal of structure, cladding and edge barriers. It may be approached by considering the probability and consequences of failure of individual structural components. Risk is defined as the product of the probability of failure and its consequences.

D2.2  Probability of failure may be affected typically by:

- current structural capacity;
- mode of failure;
- critical loading;
- structural function of the component;
- material properties and dimensions;
- likelihood of accidental loading, e.g. vehicle impact;
- sensitivity of detail to workmanship and deterioration.

D2.3 Consequences of failure may be judged typically by consideration of:

- potential loss of life and injury of people;
- loss of business;
- costs of repair or reconstruction;
- loss of confidence.

D2.4 Factors affecting probability and consequences differ in their importance. Such differences can be taken into consideration when determining relative risks. Owners and Operators need to be involved appropriately in consideration of the options for action following assessments of risks.

## D3 Appraisal of remaining capacity of structural components

D3.1 A key consideration determining appropriate options for the future of the structure is the remaining load-carrying capacity and residual life and its decay over time if nothing is done, or if action is taken to slow deterioration processes.

D3.2 Current remaining capacity of structural components should be based on the findings of the Condition Survey and Structural Investigation. Essentially remaining capacity of concrete components may be estimated using 'safe-side' assumptions for loss of reinforcement cross-section, and for effective anchorage of reinforcement and concrete cross-sectional area due to delamination and disruption of the concrete. Where a structure has been repaired, reliance should not be placed on the structural effectiveness of the repairs unless they can be reconciled with drawings and specifications, and have been tested and certified by an Engineer.

D3.3 Appraisal of future safety risks can be made more quantitative using estimates of current remaining capacity and predictions of the time until the component becomes unsafe and must be repaired or strengthened.

D3.4 Quantitative prediction of the future progress of deterioration and condition is only possible using a calibrated model that gives quantitative data on changes over time for the specific site conditions. Owing to the variable nature of the environment and the quality of the construction, a structure-specific, calibrated model based on actual test data will be needed. For reinforced concrete, deterioration models are available that take into account the primary cause of deterioration and the prediction mechanism that affect the integrity of the component.[D1–D3] To enable prediction of the strength of components in the future where reinforcement corrosion is expected, models usually give estimates of the time to the start of corrosion, the subsequent rate of corrosion and the amount of corrosion to cause delamination of the concrete cover. It is then possible to determine the degraded cross-section sizes and hence, by structural analysis, to determine the changes in structural actions and residual capacities at any future time. A series of calculations for different future times can enable an estimate of the time by which repair or strengthening will be required to maintain the safety of the structure.

D3.5 The accuracy of predictions for the rates of corrosion deterioration can be confirmed or adjusted as time passes by the use of monitoring at selected points on the structure.

D3.6 A quantitative approach may allow a reinforced concrete structure to remain in use whilst allowing its deterioration to be managed; for example, removal of delaminated concrete to avoid the hazard of falling spalls. The timing of repair may then be optimized taking into consideration commercial and operational parameters whilst ensuring that structural integrity is not compromised.

D3.7    An example of the application of an empirical corrosion model, developed for use in bridge management based on whole-life assessment of bridges, shows what can be achieved with historic data.[D4] Application to a population of concrete car park structures may be beneficial, but development for this use is first needed, taking into account the inherent variability of materials and environment.

D3.8    For reliable application, such quantitative predictions require substantial experience of the structural deterioration processes involved in concrete structures. More specialized techniques of probabilistic assessment using structural reliability analysis are also available to determine failure probabilities.[D5]

## D4  Appraisal of capacity and ductility of sensitive structural concrete details

D4.1    Sensitive structural details are those that would fail structurally as a result of relatively small loss of structural material strength, or of relatively small change in the forces (direction of action and/or magnitude) transferred through the detail. Details that may be sensitive include corbels supporting reinforced concrete beams, columns supported on transfer beams or deck edges, and columns supporting reinforced concrete slabs, especially where there is a hole in the slab adjacent to the column for services.

D4.2    In assessing the capacity of such details, it is important to recognize that BS 8110 does not allow for any strength loss in reinforced concrete from deterioration. For concrete column/slab structures, slab support at columns is generally structurally sensitive. Bottom slab reinforcement generally passes around columns rather than through them, with the result that a shear failure of the slab support may be sudden rather than ductile, especially if the materials in the support region are of poor quality or deteriorated. Consequently, once concrete surface deterioration starts to reduce the effective anchorage of top reinforcement around a punching shear perimeter, a structure should be regarded as at risk. This situation may be judged to exist if the concrete has degraded to the depth of the steel.

D4.3    In reinforced concrete structures, ductility of behaviour under bending loads is ensured generally by the provision of properly anchored tension steel reinforcement in beams and slabs. For ductile behaviour of such components in shear, shear stirrup reinforcement is used and bottom steel reinforcement is provided across potential shear surfaces anchored into concrete on either side. In concrete column/slab structures, ductile behaviour in punching shear at slab supports is most reliably ensured by detailing the bottom slab steel directly through columns. This form of detailing has not generally been adopted in concrete car park structures for construction practice reasons and because the design codes of practice CP114, CP110 and BS 8110 do not insist on it. The details used at column/slab supports have been evaluated by some tests that have demonstrated that some degree of punching shear ductility is present. However, loss of the limited as-built ductility should be assumed in Structural Appraisal to result potentially in a sudden punching shear failure mode if significant deterioration is present. The shear strength of the slab support decreases as the depth of friable slab concrete or the splitting of cover by reinforcement corrosion relaxes the bond to reinforcement. The shear strength should therefore be appraised both as-built and as-deteriorated.[D6] Where repairs have been undertaken previously, their structural effects should be considered, including damage from cutting out, the effectiveness of propping during repair works, the risks of repair debonding and measures to enhance punching strength.

D4.4    Structural Appraisal should consider the potential for punching shear failure to trigger a progressive collapse. Redistribution of reactions and moments allowed in the design of insitu reinforced concrete structures should not be assumed in Structural Appraisal where structural elements might have sudden failure modes arising from the nature of the reinforcement detailing or from degradation of the concrete (see also D6).

## D5  Structural sensitivities of early reinforced concrete slab structures, including Lift-Slab structures

D5.1    Early insitu reinforced concrete column and slab structures were usually designed to CP114. Its provisions for design against punching shear at column/slab connections give a poor, generally optimistic, prediction of strength. Consequently, the margin of safety in these structures against punching shear is generally lower than for structures of more modern design, i.e. those designed to BS 8110. Both BS 8110: 1985 and BS 8110: 1997 include a substantially improved method for punching shear design in normal insitu flat slab construction with regular spans, but give little guidance on structures where spans are irregular. For this reason, and also because of other aspects on which the code does not give guidance, e.g. uneven spacing of reinforcement near to columns, holes adjacent to columns and higher shear stresses arising on the weaker face of the shear perimeter, substantial ambiguities arise in interpreting BS 8110: 1997 for Structural Appraisal purposes. For Lift-Slab structures there is uncertainty also in how to take into account, for example, stress concentration from the wedge supported shear head and the situation where the angles supporting the slab are positioned above the slab soffit.[D6]

D5.2    It should be borne in mind, when using BS 8110 as a basis for the Structural Appraisal of column/slab structures, that the design procedures might not achieve levels of safety against punching shear comparable to those for bending. Caution is also needed when using some structural analysis methods to calculate the effective shear on the punching perimeter. Effective shear may be underestimated.

D5.3    Some early concrete column and slab structures may be found where enhancement of punching shear strength at connections is needed or where enhancement works have already been completed. Techniques that may have been used include use of column heads, brackets to form a capital at the top of the column, pressure grouting with epoxy grout, or back-propping to restore the continuity moment and then breaking out and recasting the connection with additional correctly anchored reinforcement.

D5.4    A few concrete column and slab car park structures were built using the Lift-Slab system of construction. This system involves the casting of the concrete floor slabs at ground level, lifting them up precast columns to the required height and then supporting them on the columns using steel wedges engaging in welded angle shear collars cast into the slab. Although four wedges are used to support the slab at each column, the reactions carried on the wedges are usually unequal and, in some cases, all the reaction may be carried on one wedge. In these circumstances the margin against punching shear failure is much reduced, especially if the concrete is of poor quality, deteriorated or repaired.[D6] Inspection of these collars is difficult, possibly requiring use of specialist non-destructive testing (NDT) techniques such as X-ray.

## D6 Evaluation of vulnerability to progressive collapse

D6.1    Where structures have a high degree of redundancy and connections between components are strong and ductile, there is little likelihood of the collapse of an individual component. Ductile modes of local failure that develop clearly visible cracking and deformations arise as the components redistribute moments and shears to alternative load paths with some reserves of strength. There is even less likelihood of that collapse developing into a progressive collapse of a substantial part or the whole of the structure.

D6.2    Many car park structures were built before the requirements for resistance to progressive collapse were introduced in the Building Regulations following the Ronan Point collapse in 1968. These requirements applied initially only to buildings of five or more storeys. Similar requirements were introduced in 1985 for buildings with spans in excess of 9 m. These latter requirements were subsequently removed from the Regulations in 1995. Consequently some car park structures were designed without explicitly meeting code provisions for resistance against progressive collapse (robustness).

D6.3    Some concrete car park structures may not therefore be inherently robust with ductile failure modes as built, especially where construction defects have exacerbated vulnerability to progressive collapse. In some cases deterioration of structural components may have led to loss of ductility over time and enhanced vulnerability and risk of development of a mechanism following local structural failure. The possible presence of these circumstances should be considered in the Structural Appraisal of any concrete car park structure.

D6.4    Consideration of the vulnerability of a car park structure to progressive collapse should be based on critical examination of the layout and form of the structure. The examination should seek to identify locations in the structure that may be susceptible to local damage or failure due to accident, e.g. vehicle impact or deterioration, and where the local damage may propagate through the structure causing more widespread collapse. Potential locations include columns exposed to a high risk of vehicle impact, especially if supporting transfer beams, or columns supported on the edges of deck slabs. Slab supports at columns of insitu reinforced or precast reinforced construction (including Lift-Slab), especially where reinforcement corrosion and concrete deterioration are present, are also potential locations for local structural failure that may lead to a progressive collapse.[D6] In identifying susceptibility to local structural damage or failure, it is helpful to seek critical circumstances through asking a series of 'What if?' questions. Each critical circumstance identified should then be evaluated for its structural consequences assuming a somewhat onerous degree of local damage or failure, e.g. complete removal of column/support.

D6.5    Consideration of the possibility of local deterioration/failure causing a collapse that propagates progressively through part or all of the structure should take into account structural defects or deterioration present in neighbouring components and connections.

Where the calculated strength of column/slab connection against punching shear failure for the undamaged state meets current Code requirements, the issue then is whether deterioration of the bond and anchorage of the top steel in the slab has weakened the connection significantly. If the connection is found to be below strength in the 'undamaged' condition and/or 'as-deteriorated', then prompt effective structural repair will be needed. Explicit consideration of the risk of progressive collapse should establish the effects of transfer of load after the failure of individual supports. For Lift-Slab and other structures sensitive to construction misfit, the construction tolerances on setting of wedges or supports should be considered in all load combinations.

## D7 References to Appendix D

D1. CONTECVET: *a validated user's manual for assessing the residual service life of concrete structures: Manual for assessing concrete structures affected by ASR; Manual for assessing concrete structures affected by frost; Manual for assessing corrosion-affected concrete structures.* EC Innovation Programme, IN309021, British Cement Association, 2002.

D2. Roberts, M. B. and Atkins, C. P. Deterioration modelling. *Concrete Engineering International*, Vol. 1, November/December 1997.

D3. Beamish S. *Proposed research on structural behaviour and material deterioration. Part 2: Material deterioration.* Conference on Interim Guidance on the Inspection and Maintenance of Multi-storey Car Parks, Institution of Civil Engineers, London, July 2000.

D4. Roberts, M. B. et al. A proposed empirical corrosion model for reinforced concrete. *Proceedings, Institution of Civil Engineers, Structures and Buildings*, February, 2000, 140, pp. 1–11.

D5. Sarja, A. and Vesikari, E. *Durability design of concrete structures.* Report 14 of RILEM Technical Committee 130-CSL.

D6. *Pipers Row car park, Wolverhampton: quantitative study of the causes of the partial collapse on 20 March 1997.* Report, Health and Safety Executive (in preparation).

# Index

Lightning Source UK Ltd.
Milton Keynes UK
UKOW05f0116180716

278521UK00004B/19/P